Food Processing Technology

Food Processing Technology

Travis Summers

STATES
ACADEMIC PRESS
www.statesacademicpress.com

States Academic Press,
109 South 5th Street,
Brooklyn, NY 11249, USA

Visit us on the World Wide Web at:
www.statesacademicpress.com

ISBN: 978-1-63989-207-5 (Hardback)

Cataloging-in-Publication Data

Food processing technology / Travis Summers.
p. cm.
Includes bibliographical references and index.
ISBN 978-1-63989-207-5
1. Food industry and trade. 2. Food industry and trade--Appropriate technology.
3. Food. I. Summers, Travis.

TP370 .F66 2022
664--dc23

Table of Contents

Preface

Food processing is a value addition process which transforms raw products related to dairy, agriculture, animal husbandry, meat, fishing or poultry. The finished product has commercial value and is suitable for human or animal consumption. Food processing can be classified into primary, secondary and tertiary food processing. Primary processes turn agricultural products into consumables. Secondary processes refer to cooking techniques, and tertiary methods are concerned with the industrial processes which are involved in production of processed food. Food processing also involves preservation processes which help in preventing oxidization of fats which in turn makes the food rancid. This book attempts to understand the multiple branches that fall under the discipline of food processing and how such concepts have practical applications. It is a compilation of chapters that discuss the most vital concepts in the field of food processing. The book is appropriate for those seeking detailed information in this area.

To facilitate a deeper understanding of the contents of this book a short introduction of every chapter is written below:

Chapter 1- The process of transformation of one form of food into other forms is known as food processing. Some of the significant aspects of food processing are retort processing of food, food deterioration and control, and food processing and preservation. This is an introductory chapter which will briefly introduce all the significant aspects of food processing.

Chapter 2- The process of adding chemicals to food products to prevent the deterioration of food caused by the growth of micro-organisms is known as chemical preservation. Some of the chemical preservatives are sodium benzoate, nitrites and nitrates, sulfites, potassium sorbate and spices. This chapter has been carefully written to provide an easy understanding of the varied facets of chemical preservation in food processing.

Chapter 3- The complicated food processes can be broken down into a small number of basic operations, which are known as unit operations. There are various processes involved in food processing such as fluid flow, heat transfer, drying, evaporation and mechanical separations. The aim of this chapter is to explore these various processes related to food processing.

Chapter 4- Some of the important food preservation techniques are curing, cooling, freezing, boiling, heating, pickling, sugaring, canning, etc. Curing is the process of removing moisture from the food by adding salt, sugar or nitrate and facilitating osmosis. The topics elaborated in this chapter will help in gaining a better perspective about these food preservation techniques.

Chapter 5- The technique of concentrating and separating particles without using heat is known as membrane technology. It includes technologies such as reverse osmosis, ultrafiltration, microfiltration and nanofiltration. This chapter closely examines the key concepts of membrane technology to provide an extensive understanding of the subject.

I owe the completion of this book to the never-ending support of my family, who supported me throughout the project.

Travis Summers

Chapter 1

Introduction to Food Processing

The process of transformation of one form of food into other forms is known as food processing. Some of the significant aspects of food processing are retort processing of food, food deterioration and control, and food processing and preservation. This is an introductory chapter which will briefly introduce all the significant aspects of food processing.

Food processing is any of a variety of operations by which raw foodstuffs are made suitable for consumption, cooking, or storage. Food processing generally includes the basic preparation of foods, the alteration of a food product into another form (as in making preserves from fruit), and preservation and packaging techniques.

A number of food-processing innovations have even resulted in new products, such as concentrated fruit juices, freeze-dried coffee, and instant foods. Foods and food supplements have also been processed from such hitherto untapped sources as oilseeds (chiefly protein-rich soybeans and cottonseeds); mutant varieties of crops; leaves, grasses, and aquatic plants; and highly nutritious fish meal and concentrates.

Food processing dates back to prehistoric ages, with crude processing methods that included slaughtering, fermenting, sun drying, preserving with salt, and various means of cooking (such as roasting, smoking, steaming, and oven baking). Salt-preservation was especially common for foods that constituted the diets of warriors and sailors, up until the introduction of canning methods.

Canning

The process of canning is sometimes called sterilization because the heat treatment of the food eliminates all microorganisms that can spoil the food and those that are harmful to humans, including directly pathogenic bacteria and those that produce lethal toxins. Most commercial canning operations are based on the principle that bacteria destruction increases tenfold for each 10° C increase in temperature. Food exposed to high temperatures for only minutes or seconds retains more of its natural flavor. In the Flash 18 process, a continuous system, the food is flash-sterilized in a pressurized chamber to prevent the superheated food from boiling while it is placed in containers. Further sterilizing is not required.

Freezing

Although prehistoric human's stored meat in ice caves, the food-freezing industry is more recent in origin than the canning industry. The freezing process was used commercially for the first time in 1842, but large-scale food preservation by freezing began in the late 19th century with the advent of mechanical refrigeration.

Freezing preserves food by preventing microorganisms from multiplying. Because the process does not kill all types of bacteria, however, those that survive reanimate in thawing food and often grow more rapidly than before freezing. Enzymes in the frozen state remain active, although at a reduced rate. Vegetables are blanched or heated in preparation for freezing to ensure enzyme inactivity and thus to avoid degradation of flavor. Blanching has also been proposed for fish, in order to kill cold-adapted bacteria on their outer surface. In the freezing of meats various methods are used depending on the type of meat and the cut. Pork is frozen soon after butchering, but beef is hung in a cooler for several days to tenderize the meat before freezing.

Frozen foods have the advantage of resembling the fresh product more closely than the same food preserved by other techniques. Frozen foods also undergo some changes, however. Freezing causes the water in food to expand and tends to disrupt the cell structure by forming ice crystals. In quick-freezing the ice crystals are smaller, producing less cell damage than in the slowly frozen product. The quality of the product, however, may depend more on the rapidity with which the food is prepared and stored in the freezer than on the rate at which it is frozen. Some solid foods that are frozen slowly, such as fish, may, upon thawing, show a loss of liquid called drip; some liquid foods that are frozen slowly, such as egg yolk, may become coagulated. Because of the high cost of refrigeration, frozen food is comparatively expensive to produce and distribute. High quality is a required feature of frozen food to justify the added cost in the market. This method of preservation is the one most widely used for a great variety of foods.

Drying and Dehydration

Although both these terms are applied to the removal of water from food, to the food technologist drying refers to drying by natural means, such as spreading fruit on racks in the sun, and dehydration designates drying by artificial means, such as a blast of hot air. In freeze-drying a high vacuum is maintained in a special cabinet containing frozen food until most of the moisture has sublimed. Removal of water offers excellent protection against the most common causes of food spoilage. Microorganisms cannot grow in a water-free environment, enzyme activity is absent, and most chemical reactions are greatly retarded. This last characteristic makes dehydration preferable to canning if the product is to be stored at a high temperature. In order to achieve such protection, practically all the water must be removed. The food then must be packaged in a moisture-proof container to prevent it from absorbing water from the air.

Vegetables, fruits, meat, fish, and some other foods, the moisture content of which averages as high as 80 percent, may be dried to one-fifth of the original weight and about one-half of the original volume. The disadvantages of this method of preservation include the time and labor involved in rehydrating the food before eating. Further because it absorbs only about two-thirds of its original water content, the dried product tends to have a texture that is tough and chewy.

Drying was used by prehistoric humans to preserve many foods. Large quantities of fruits such as figs have been dried from ancient times to the present day. In the case of meat and fish, other preservation methods, such as smoking or salting, which yielded a palatable product, were generally preferred. Commercial dehydration of vegetables was initiated in the United States during the American Civil War but, as a result of the poor quality of the product, the industry declined sharply after the war. This cycle was repeated with subsequent wars, but after World War II the dehydration industry thrived. This industry is confined largely to the production of a few dried foods, however, such as milk, soup, eggs, yeast, and powdered coffee, which are particularly suited to the dehydration method. Present-day dehydration techniques include the application of a stream of warm air to vegetables. Protein foods such as meat are of good quality only if freeze-dried. Liquid food is dehydrated usually by spraying it as fine droplets into a chamber of hot air, or occasionally by pouring it over a drum internally heated by steam.

Freeze-drying

A processing method that uses a combination of freezing and dehydration is called freeze-drying. Foods that already have been frozen are placed in a vacuum-tight enclosure and dehydrated under vacuum conditions with careful application of heat. Normally ice melts and becomes water when heat is applied. If more heat is applied, it turns to steam. But in freeze-drying, the ice turns directly to vapor, and there is little chance that microorganisms will grow. Freeze-dried foods, like those that are dehydrated, are light and require little space for storage and transportation. They do not need to be refrigerated, but they must be reconstituted with water before they are ready to consume.

Irradiation

As early as 1895, a major breakthrough in the world of science had arisen; the discovery of the X-ray by German physicist Wilhelm von Roetengen. This technological advancement, along with the soon to be discovered concept of radioactivity by French physicist Antoine Henri Becquerel, became the focus of attention for many scientifically based studies. Of most importance, to the field of food preservation, these two discoveries began the now controversial process of food irradiation.

Food irradiation employs an energy form termed ionizing radiation. In short, this process exposes food particles to alpha, beta and/or gamma rays. The rays cause whatever material they strike to produce electrically charged particles called ions. Ionizing radiation provides many attributes to treating foods. It has the ability to penetrate deeply into a food interacting with its atoms and molecules, and causing some chemical and biological effects that could possibly decrease its rate of decay. It also has the ability to sanitize foods by destroying contaminants such as bacteria, yeasts, molds, parasites and insects. Irradiation delays ripening of fruits and vegetables; inhibits sprouting in bulbs and tubers; disinfests grain, cereal products, fresh and dried fruits, and vegetables of insects; and destroys bacteria in fresh meats. The irradiation of fresh fruits and vegetables, herbs and spices, and pork was approved in 1986. In 1990 the FDA approved irradiation of poultry to control salmonella and other disease-causing microorganisms. Irradiated foods were used by U.S. astronauts and by Soviet cosmonauts. Public concern over the safety of irradiation, however, has limited its full-scale use. It is still off to a slow start, with only one food irradiation plant open in Mulberry, Florida, but it is seemingly catching the eyes of the producers and the consumers throughout the world.

Miscellaneous Methods

Other methods or a combination of methods may be used to preserve foods. Salting of fish and pork has long been practiced, using either dry salt or brine. Salt enters the tissue and, in effect binds the water, thus inhibiting the bacteria that cause spoilage. Another widely used method is smoking, which frequently is applied to preserve fish, ham, and sausage. The smoke is obtained by burning hickory or a similar wood under low draft. In this case some preservative action is provided by such chemicals in the smoke as formaldehyde and creosote, and by the dehydration that occurs in the smokehouse. Smoking usually is intended to flavor the product as well as to preserve it.

Sugar a major ingredient of jams and jellies, is another preservative agent. For effective preservation the total sugar content should make up at least 65 percent of the weight of the final product. Sugar, which acts in much the same way as salt, inhibits bacterial growth after the product has been heated. Because of its high acidity, vinegar (acetic acid) acts as a preservative. Fermentation caused by certain bacteria, which produce lactic acid, is the basis of preservation in sauerkraut and fermented sausage. Sodium benzoate, restricted to concentrations of not more than 0.1 percent, is used in fruit products to protect against yeasts and molds. Sulfur dioxide, another chemical preservative permitted in most states, helps to retain the color of dehydrated foods. Calcium propionate may be added to baked goods to inhibit mold.

Packaging

The packaging of processed foods is just as important as the process itself. If foods are not packaged in containers that protect them from air and moisture, they are subject to spoilage. Packaging materials must therefore be strong enough to withstand the heat and cold of processing and the wear and tear of handling and transportation.

From the time the canning process was developed in the early 19th century until the beginning of the 20th century, cans and glass containers were the only packages used. The first cans were crude containers having a hole in the top through which the food was inserted. The holes were then sealed with hot metal. All cans were made by hand from sheets of metal cut to specific sizes. In about 1900 the sanitary can was invented. In this process, machines form cans with airtight seams. A processor buys cans with one end open and seals them after filling. Some cans are made of steel coated with tin and are often glazed on the inside to prevent discoloration. Some are made of aluminum. Frozen foods are packaged in containers made of layers of fibreboard and plastic or of strong plastic called polyethylene. Freeze-dried and dehydrated foods are packed in glass, fibreboard, or cans.

Retort Processing of Food

A Retort is a pressure vessel used in the food manufacturing industry to "commercially sterilize" food after it has been placed into its container and the container has been hermetically sealed. The method of sterilization is a thermal based process, or the application of high heat to the product. The high temperatures required in a sterilization process destroys pathogenic organisms that may be present in/on the container and/or food product, and is well above the boiling point of water at normal atmospheric pressure. So, the retort is by design a "pressure vessel". Thermal processing/sterilization of shelf stable, low acid foods (i.e.: defined as products with a finished equilibrium pH greater than 4.6 and a water activity greater than 0.85) is usually performed at temperatures at or above 250°F or 121°C, give or take ~10°F/5.6°C. The higher the temperature, the shorter the time the product needs to be exposed to heat.

Batch retorts are available in several configurations. First, they can be static or agitating style machines. A static retort is typically used for conductive (i.e.: viscous) heating products where the product does not benefit from any kind of movement during the process (heat or cooling). Agitating style retorts are typically used for convective (i.e.: "flow able liquid") type products that benefit from some product movement (i.e.: "stirring") in the container during the process. Either from a process stand-point, to improve the rate of heat transfer into/out of the container, and/or from a product quality stand-point (i.e.: to shorten the exposure time to heat). Agitating style retorts can utilize various methods of agitation depending on the orientation of the product container. Vertically oriented containers, such as cans or glass/plastic bottles, are typically agitated in a rotary fashion, either axially or end-over-end. With horizontally oriented containers, such as flexible pouches and semi-rigid bowls and/or trays (which contain product that benefits from agitation) the preferred method is horizontal reciprocation.

Batch retorts can employ various methods of process delivery. Some of which also utilize overpressure or counter-pressure to help protect the integrity of the container during the process (i.e.: to keep the package from bursting as temperature and pressure builds inside the container during the process). Rigid containers, such as steel cans, can withstand large differences between the pressure inside and outside of the container, and therefore these types of containers typically do not require overpressure. They can be processed in a 100% saturated steam environment without the use of overpressure during the heating phases. On the other hand, more fragile flexible and semi-rigid containers cannot withstand high pressure differentials, so air is introduced into the retort to provide overpressure to maintain package integrity during the process. These types of containers require more sophisticated overpressure process delivery methods such as water spray, water cascade or water shower, water immersion or steam-air type systems. Because air is an insulator, a means of stirring or mixing the process media in the retort is required to avoid cold spots in the machine, thus ensuring good temperature distribution throughout the retort and product load. This mixing is accomplished by the different water flow methodologies, or via a fan in the case of steam-air retorts, and/or via mechanical rotation of the insert/drum in the case of agitating style machines.

Overpressure is also important in the cooling phases of a retort process because as cooling water is introduced into the retort it collapses the steam created in the heating step(s). Without sufficient introduction of air overpressure during cooling, the pressure in the retort can suddenly drop due to steam collapse thus creating a vacuum situation in the retort. If this happens the pressure differential between the outside environment and the temperature/pressure environment inside the container becomes too great thus causing the container to burst (otherwise known as "buckling"). Precise control of overpressure during the initial phases of cooling is important to avoid the above situation but ramping that pressure down in the latter stages of cooling is important as well to avoid crushing the container (or otherwise known as "panelling") as the temperature and pressure inside the container subsides. While the retort process inactivates or destroys bacterial pathogens, it does not destroy all microscopic spoilage organisms. Thermophiles are bacteria that can withstand temperatures well above typical retort temperatures. For this reason, the product must be cooled down to a temperature below that at which these organisms will reproduce, thus causing thermophilic spoilage.

Food Processing and Preservation

The main reasons for food preservation are to overcome inappropriate planning in agriculture, produce value-added products, and provide variation in diet. The agricultural industry produces raw food materials in different sectors. Inadequate management or improper planning in agricultural production can be overcome by avoiding inappropriate areas, times, and amounts of raw food materials as well as by increasing storage life using simple methods of preservation. Value-added food products can give better-quality foods in terms of improved nutritional, functional, convenience, and sensory properties. In food preservation, the important points that need to be considered are:

- The desired level of quality.
- The preservation length.
- The group for whom the products are preserved.

Preservation of Foods

After storage of a preserved food for a certain period, one or more of its quality attributes may reach an undesirable state. Quality is an illusive, ever changing concept. In general, it is defined as the degree of fitness for use or the condition indicated by the satisfaction level of consumers. When food has deteriorated to such an extent that it is considered unsuitable for consumption, it is said to have reached the end of its shelf life. The product quality attributes can be quite varied, such as appearance, sensory, or microbial characteristics. Loss of quality is highly dependent on types of food and composition, formulation (for manufactured foods), packaging, and storage conditions. Quality loss can be minimized at any stage of food harvesting, processing, distribution, and storage. The product quality can be defined using many factors, including appearance, yield, eating characteristics, and microbial characteristics, but ultimately the final use must provide a pleasurable experience for the consumer. The various stages of food production, manufacture, storage, distribution, and sale are shown in Figure.

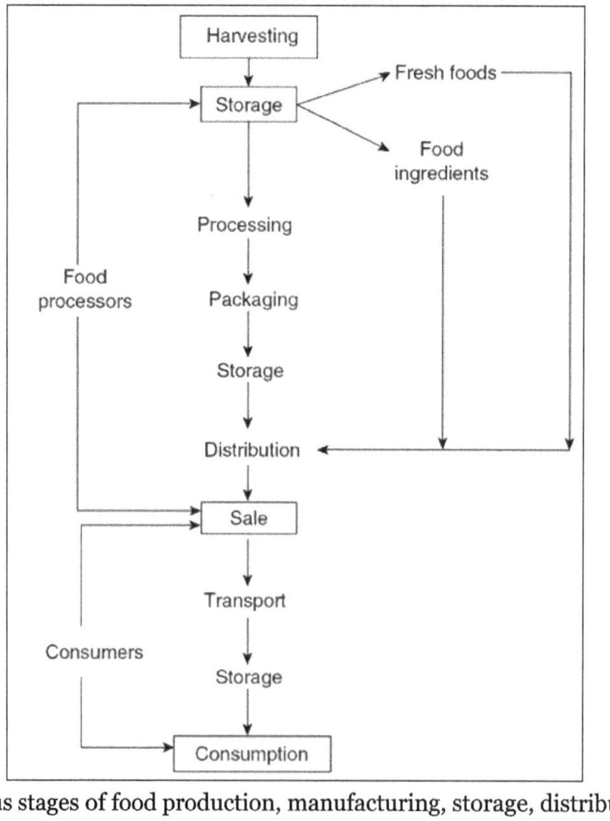

Figure: Various stages of food production, manufacturing, storage, distribution & sale.

Quality loss can be minimized at any stage and thus quality depends on the overall control of the processing chain. The major quality-loss mechanisms and consequences are shown in table and figure.

Microbiological	Enzymatic	Chemical	Physical	Mechanical
Microorganism growth	Browning	Color loss	Collapse	Bruising due to vibration
Off-flavor	color change	Flavor loss	Controlled release	Cracking
Toxin production	Off-flavor	Nonenzymatic browning	Crystallization	Damage due to pressure
		Nutrient loss	Flavor encapsulation	
		Oxidation-reduction	Phase changes Recrystallization shrinkage Transport of component	

Table: Major quality loss mechanisms.

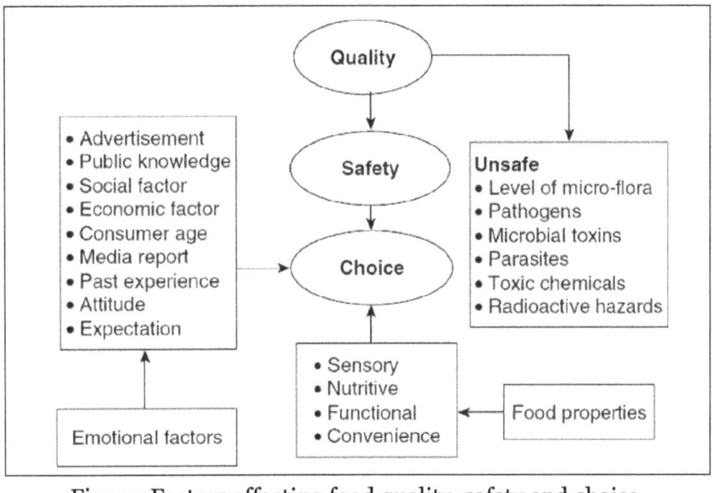

Figure: Factors affecting food quality, safety and choice.

Causes of Deterioration

Mechanical, physical, chemical, and microbial effects are the leading causes of food deterioration and spoilage. Damage can start at the initial point by mishandling of foods during harvesting, processing, and distribution; this may lead to ultimate reduction of shelf life. Other examples of deterioration can be listed as follows: (i) bruising of fruits and vegetables during harvesting and postharvest handing, leading to the development of rot, (ii) tuberous and leafy vegetables lose water when kept in atmospheres with low humidity and, subsequently, wilt, and (iii) dried foods kept in high humidity may pick up moisture and become soggy. The four sources of microbial contaminants are soil, water, air, and animals (insects, rodents, and humans). In preservation, each component factor needs to be controlled or maintained to a desired level.

During storage and distribution, foods are exposed to a wide range of environmental conditions. Environmental factors such as pressure, temperature, humidity, oxygen, and light can trigger several reactions that may lead to food degradation. Mechanical damage (e.g., bruises and wounds) is conducive to spoilage, and it frequently causes further chemical and microbial deterioration.

Table: Organisms that spoil food.

Microorganisms
Fungi: mold and yeast
Bacteria
Phages
Protozoa
Insects and mites
Directly by eating (infestation)
Indirectly by spreading diseases (fruitfly, housefly)
Rodent
Directly by consuming food
b. Indirectly by spreading diseases

Peels, skins, and shells constitute natural protection against this kind of spoilage. Shriveling occurs due to the loss of water from harvested fruits and vegetables. Each microorganism has (i) an optimum temperature at which it grows best, (ii) a minimum temperature below which growth no longer takes place, and (iii) a maximum temperature above which all development is suppressed. Bacteria that grow particularly well at low temperatures are called psychrophilic (cryophilic) or low-temperature organisms. Bacteria with an optimum temperature of 20°C–45°C are mesophilic, and those with an optimum temperature above 45°C are thermophilic. Microbial growth in foods results in food spoilage with the development of undesirable sensory characteristics, and in certain cases the food may become unsafe for consumption. Microorganisms have the ability to multiply at high rates when favorable conditions are present. Prior to harvest, fruits and vegetables generally have good defense mechanisms against microbial attack; however, after separation from the plant, they can easily succumb to microbial proliferation.

The presence of unsaturated fatty acids in foods is a prime reason for the development of rancidity during storage as long as oxygen is available. While development of off-flavors is markedly noticeable in rancid foods, the generation of free radicals during the autocatalytic process leads to other undesirable reactions, for example, loss of vitamins, alteration of color, and degradation of proteins. The presence of oxygen in the immediate vicinity of food leads to increased rates of oxidation. Similarly, water plays an important role; lipid oxidation occurs at high rates at very low water activities. Some chemical reactions are induced by light, such as loss of vitamins and browning of meats. No enzymatic browning is a major cause of quality change and degradation of the nutritional content of many foods. This type of browning reaction occurs due to the interaction between reducing sugars and amino acids, resulting in the loss of protein solubility, darkening of lightly colored dried products, and development of bitter flavours. Environmental factors such as temperature, water activity, and pH have an influence on no enzymatic browning.

Principles of Food Preservation

- Slowing down or inhibiting chemical deterioration and microbial growth.
- Directly inactivating bacteria, yeasts, molds, or enzymes.
- Avoiding recontamination before and after processing.

A number of techniques or methods from the above categories are shown in figure:

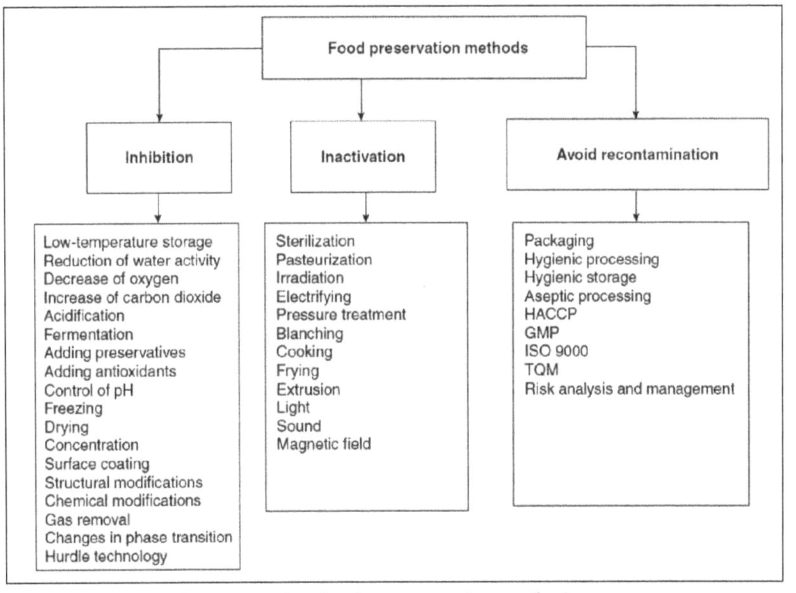

Figure: Major food preservation techniques.

A good method of food preservation is one that slows down or prevents altogether the action of the agents of spoilage. Also, during the process of food preservation, the food should not be damaged. In order to achieve this, certain basic methods were applied on different types of foods. For example in earlier days, in very cold weather condition, ice was used to preserve foods. Thus, very low temperature became an efficient method for preventing food spoilage. The principles of food preservation are:

- Removal of micro-organisms or inactivating them: This is done by removing air, water (moisture), lowering or increasing temperature, increasing the concentration of salt or sugar or acid in foods. If you want to preserve green leafy vegetables, you have to remove the water from the leaves so that microorganisms cannot survive. You do this by drying the green leaves till all the moisture evaporates.
- Inactivating enzymes: Enzymes found in foods can be inactivated by changing their conditions such as temperature and moisture, when you preserve peas, one of the methods of preservations is to put them for a few minutes in boiling water. This method also known as blanching inactivates enzymes and thus, helps in preserving the food.
- Removal of insects, worms and rats: By storing foods in dry, air tight containers the insects, worms or rats are prevented from destroying it.

While the currently used preservation procedures continue in one or more of these three ways, there have recently been great efforts to improve the quality of food products principally to meet the requirements of consumers through the avoidance of extreme use of any single technique. Preservation starts when the harvested foods are separated from the medium of immediate growth (plant, soil, or water) or meat from the animal after slaughter, or milk from normal secretion of mammalian glands. Postharvest technology is concerned with handling, preservation, and storage of harvested foods, and maintaining its original integrity, freshness, and quality. The methods of preservation depend on the origin of foods particularly whether they are of plant or animal origin. Postharvest handling of foods of plant origin includes efficient control of environmental atmosphere, such as humidity, gas composition, and temperature, and implementing an adequate packing, storage, and transport system. Physical treatments usually used are curing, pre-cooling, temperature treatments, cleaning, and waxing, whereas chemical treatments are disinfection, fumigation, and dipping. Chemical disinfectants vary in their ability to kill microorganisms. Several chemicals are utilized, such as chlorine, chlorine dioxide, hydrogen peroxide, ozone, peroxyacetic acid, bromine, iodine, trisodium phosphate, and quaternary ammonium compounds. Although fumigants are not strictly preservatives, they are used for insect control. Methyl bromide is one of the fumigants used, but it has the potential to damage atmospheric ozone and is being phased out. There is a need for development of new environmentally safe methods of fumigation.

Inhibition

The methods based on inhibition include those that rely on control of the environment (e.g., temperature control), those that result from particular methods of processing (e.g., microstructural control), and those that depend on the intrinsic properties built into particular foods (e.g., control by the adjustment of water activity or pH value. The danger zone for microbial growth is considered to be between 5°C and 60°C; thus chilling and storing at a temperature below 5°C is one of the most popular methods of food preservation.

Use of Chemicals

The use of chemicals in foods is a well-known method of food preservation. Wide varieties of chemicals or additives are used in food preservations to control pH, as ant microbes and antioxidants, and to provide food functionality as well as preservation action. Some additives are entirely synthetic (not found in nature), such as phenolic antioxidant tertiary butyl hydroquinone (TBHQ), and others are extracted from natural sources, such as vitamin E. Irrespective of origin, food additives must accomplish some desired function in the food to which they are added, and they must be safe to consume under the intended conditions of use. Many legally permitted preservatives in foods are organic acids and esters, including sulfites, nitrites, acetic acid, citric acid, lactic acid, sorbic acid, benzoic acid, sodium diacetate, sodium benzoate, methyl paraben, ethyl paraben, propyl paraben, and sodium propionate. When a weak acid is dissolved in water, equilibrium is established between un dissociated acid molecules and charged anions, the proportion of un dissociated acid increasing with decreasing ph. There are several limitations to the value of organic acids as microbial inhibitors in foods:

- They are usually ineffective when initial levels of microorganisms are high.
- Many microorganisms use organic acids as metabolizable carbon sources.
- There is inherent variability in resistance of individual strains.
- The degree of resistance may also depend on the conditions.

Controls of Water and Structure

Many physical modifications are made in ingredients or foods during preservation. Such modifications can also improve the sensory, nutritional, and functional properties of foods. Changes experienced by foods during processing include glass formation, crystallization, caking, cracking, stickiness, oxidation, gelatinization, pore formation, and collapse. Through precise knowledge and understanding of such modifications, one can develop safe, high-quality foods for consumption. Water is an important constituent of all foods. The minimum water activity is the limit below which a microorganism or group of microorganisms can no longer reproduce. For most foods, this is in the water activity range of 0.6–0.7. Pathogenic bacteria cannot grow below a water activity of 0.85–0.86, whereas yeast and molds are more tolerant of a reduced water activity of 0.80, but usually no growth occurs below a water activity of about 0.62. The critical limits of water activity may also be shifted to higher or lower levels by other factors, such as pH, salt, antimicrobial agents, heat treatment, and temperature to some extent. Removing water, adding solutes, or change of solute–water interactions can reduce the water activity of a food.

Control of Atmosphere

Packaging techniques based on altered gas compositions have a long history. The respiratory activity of the various plant products generates a low-oxygen and high carbon dioxide atmosphere, which retards the ripening of fruit. Modified-atmosphere packaging is a preservation technique that may further minimize the physiological and microbial decay of perishable produce by keeping them in an atmosphere that is different from the normal composition of air. The gas composition and method of this technique depends on the types of produce and purposes. There are different ways of maintaining a modified atmosphere. In modified atmosphere packaging (termed "passive atmosphere"), the gas composition within the package is not monitored or adjusted. In "controlled atmosphere packaging," the altered gas composition inside the packaging is monitored and maintained at a pre-set level by means of scrubbers and the inlet of gases. Active packaging can provide a solution by adding materials that absorb or release a specific compound in the gas phase. Compounds that can be absorbed are carbon dioxide, oxygen, water vapour, ethylene, or volatiles that influence taste and aroma. Vacuum and modified-humidity packaging contain a changed atmosphere around the product. Although this technique was initially developed to extend the shelf life of fresh products, it is now extended to minimally processed foods from plant and animal sources, and also to dried foods.

Inactivation

1. Use of Heat Energy: Earlier, mostly heat was used for inactivation. Thermal inactivation is still the most widely used process of food preservation. The advantages of using heat for food preservation are:

- Heat is safe and chemical-free.
- It provides tender cooked flavours and taste.
- The majority of spoilage microorganisms are heat labile.
- Thermally processed foods, when packed in sterile containers, have a very long shelf life.

The main disadvantages of using heat are (i) overcooking may lead to textural disintegration and an undesirable cooked flavour, and (ii) nutritional deterioration results from high temperature processing. Heat treatment processes include mainly pasteurization, sterilization, cooking, extrusion, and frying. Recently, more electro technologies have been used and this will expand further in the future.

2. Use of High Pressure and Ultrasound: High-quality fresh foods are very popular, so consequently there is a demand for less extreme treatments and fewer additives. High-pressure hydrostatic technology gained attention for its novelty and non-thermal preservation effect. Ultrasound is sound energy with a frequency range that covers the region from the upper limit of human hearing, which is generally considered to be 20 kHz. The two applications of ultrasound in foods are (i) characterizing a food material or process, such as estimation of chemical composition, measurements of physical properties, non-destructive testing of quality attributes, and monitoring food processing, and (ii) direct use in food preservation or processing. The beneficial or deteriorative use of ultrasound depends on its chemical, mechanical, or physical effects on the process or products.

3. Use of Electricity: Many different forms of electrical energy are used in food preservation, e.g., ohmic heating, microwave heating, low electric field stimulation, high-voltage arc discharge, and high-intensity pulsed electric field. Ohmic heating is one of the earliest forms of electricity applied to food pasteurization. This method relies on the heat generated in food products as a result of electrical resistance when an electric current is passed through them. Microwave heating has been extensively applied in everyday households and the food industry, but the low penetration depth of microwaves into solid food causes thermal non uniformity. Low electric field stimulation has been explored as a method of bacterial control in meat. The plasma membranes of cells become permeable to small molecules after being exposed to an electric field; permeation then causes swelling and the eventual rupture of the cell membrane. The reversible or irreversible rupture (or electroporation) of a cell wall membrane depends on factors such as intensity of the electric field, number of pulses, and duration of pulses. This new electro-heating could be used to develop new products with diversified functionality.

4. Use of Radiation: Ionization radiation interacts with an irradiated material by transferring energy to electrons and ionizing molecules by creating positive and negative ions. The irradiation process involves exposing the foods, either pre-packaged or in bulk, to a predetermined level of ionization radiation. Irradiation has wide scope in food disinfection, shelf life extension, decontamination, and product quality improvement. Although it has high potential, there is concern on legal aspects and safety issues, and consumer attitude toward this technology. Ultraviolet (UV) radiation has long been known to be the major factor in the antibacterial action of sunlight. It is mainly used in sterilizing air and thin liquid films due to its low penetration depth. Pulsed light is a sterilization method in applications where light can access all the important volume and surfaces. Examples include packaging materials, surfaces, trans missive materials (such as air, water, and many solutions), and many pharmaceuticals or medical products. The white light pulse is generated by electrically ionizing a xenon gas-filled lamp for a few hundred millionths of a second with a high power, high-voltage pulse. In many cases, it would be very difficult to make a clear distinction between inhibition and inactivation.

5. Use of Magnetic Field: Magnetism is a phenomenon by which materials exert an attractive or repulsive force on other materials. The origin of magnetism lies in the orbital and spin motions of electrons and how the electrons interact with each other. Magnetic fields have potential in pasteurization, sterilization, and enhancing other factors beneficial to processing in food preservation.

Avoid Recontamination (Indirect Approach)

In addition to the direct approach, other measures such as packaging and quality management tools need to be implemented in the preservation process to avoid contamination or recontamination. Although these measures are not preservation techniques, they play an important role in producing high-quality safe food. With respect to the procedures that restrict the access of microorganisms to foods, the employment of aseptic packaging techniques for thermally processed foods has expanded greatly in recent years both in the numbers of applications and in the numbers of alternative techniques that are commonly available. From skins, leaves, and bark, tremendous progress has been made in the development of diversified packaging materials and equipment's. Packaging performs three main functions. The first is to control the local environmental conditions to enhance storage life. The second is the display, i.e., preservation of the product in an attractive manner to the potential buyer. The third function is to protect the product during transit to the consumer. The new concept of active or life packaging materials allows one-way transfer of gases away from the product or the absorption of gases detrimental to the product, antimicrobials in packaging, release of preservatives from controlled-release surfaces, oxygen scavengers, carbon dioxide generators, absorbers or scavengers of odours, absorption of selected wavelengths of light, and there are capabilities for controlled automatic switching. Another concept of edible or biodegradable packaging has also been evolved for environmental reasons. Processing and packaging can be integrated to improve efficiency.

References

- Food-processing: britannica.com, Retrieved 03, April 2020
- Retort-processes: retorts.com, Retrieved 19, June 2020

Chapter 2

Unit Operations and Processes

The complicated food processes can be broken down into a small number of basic operations, which are known as unit operations. There are various processes involved in food processing such as fluid flow, heat transfer, drying, evaporation and mechanical separations. The aim of this chapter is to explore these various processes related to food processing.

Food technologists apply science to convert raw materials in to healthy, safe, palatable and convenient consumable food products. They apply engineering principles in research and development, production technology, quality control, packaging and processing to develop new food products for the society. The job of a food technologist is enormous. It starts from the quality control of raw materials in the supplier's factory through the manufacturing processes till consumption of finished products. Food technologists design the processes; modify the existing processes according to the need of the raw materials interactions to processing parameters and intended quality attributes desired in the food products. So, food technologist should have adequate knowledge on the processes and unit operations adopted in food processing so that they can manipulate the system as per the requirement of the product concerned.

Process

A process may be defined as a set of activities or operations which take place in manufacturing equipment's; bring about a series of physico-chemical and biological changes in the raw food materials and as a result useful consumable products are derived. The products thus obtained are considered to be safe for consumption with greater market acceptability and enhanced storage stability, ease in handling and transportation. As for example, tomatoes are harvested from the field, the size and colour of each tomato varies from one another i.e. there is no uniformity in shape, size and colour. There may be chances of surface contamination by foreign matters like soil, leaves, pesticides etc. The tomatoes harvested from the field are not safe for use in cooking. So to make it safe we adopt some processes like cleaning, sorting, blanching etc. The marketability of tomato can be increased by sorting and grading of tomatoes before sending them to supermarkets. If we want to preserve the perishable tomato for long time, we transform them physically and chemically to make products like pickles, puree/ketchup etc. The processes involved in transformation are peeling, deseeding, crushing, pulping, heat treatment etc.

Food processing is concerned with the processes involved in the development of food products. It requires a thorough understanding about the processes, principles of operations to transform raw materials physically, chemically and biochemically in to desirable products. Moreover, the processes are manipulated by changing the parameters like machine specifications and environmental conditions inside the machine (temperature, pressure, humidity etc).

Process Flow Chart

The flow charts, also called as flow diagrams or flow sheets are the graphical representation of the processes. In short, a flow diagram shows the major operations of a process in sequence and the flow of raw materials along with the products and by-products derived from the process. The processes are represented in schematic flow charts where each processes are written inside a 'rectangle' or 'box'. Therefore, the flow diagrams are also called as block diagrams. The series of processes that indicate the manufacturing steps are connected with arrow marks representing the direction of flow of feed materials. Additional information on processes like process conditions viz. temperature and pressure etc. and flow rate of feed material may also be given with the flow diagrams. Figure shows a typical flow chart showing the food processes used during manufacture of tomato ketchup.

Another type of representation of a food process is, the equipments used for doing a particular operation are presented in pictorial view. This representation provides much detail information about a process. The standard symbolic pictures of important equipments are given in figure.

Other way of representing the process is by use of custom symbols that resembles like the original equipment. The drawing of this kind of equipment flow chat is difficult to some extent. Process piping is schematically added to the equipments kept in sequence, of course the layout of the equipments is not to the scale. An equipment flow chart of soy protein isolate manufacturing process is given in figure.

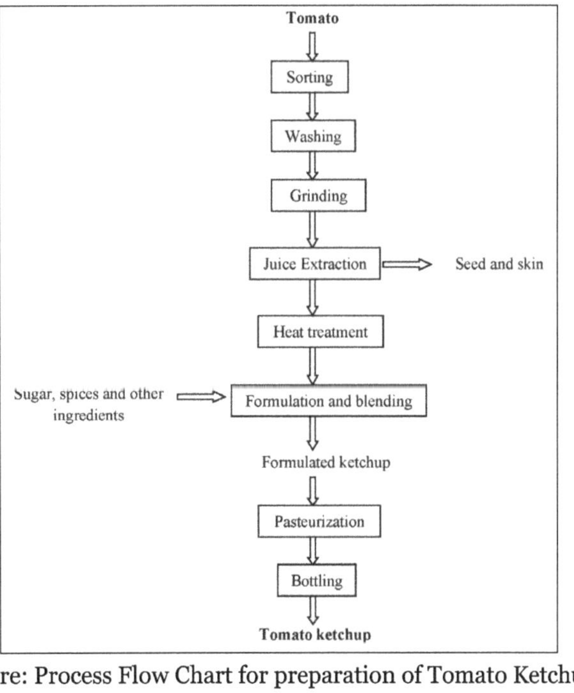

Figure: Process Flow Chart for preparation of Tomato Ketchup.

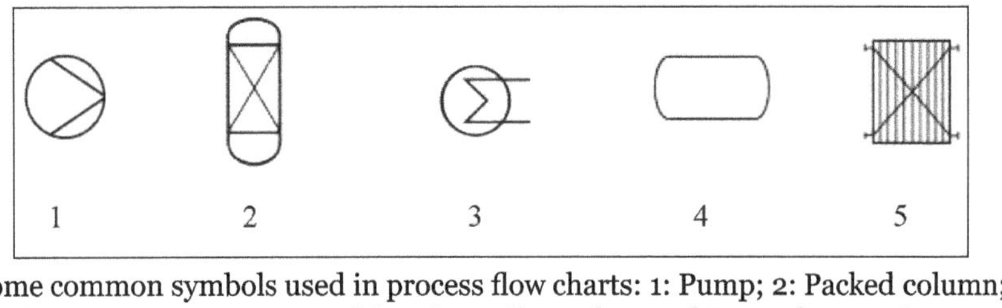

Figure: Some common symbols used in process flow charts: 1: Pump; 2: Packed column; 3: Heat exchanger; 4: Pressure vessel (Horizontal); 5: Plate and Frame heat exchanger.

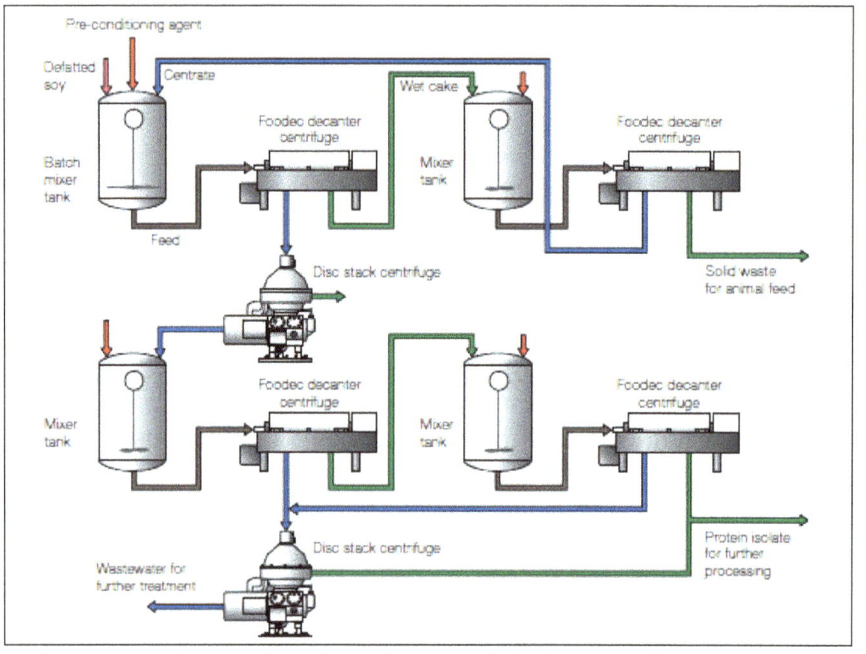

Figure: An equipment flow chart of soy protein isolate manufacturing system.

Unit Operations

A set of processes working with a common scientific principle are clumped together for ease in understanding in the study of food processing. The basic operations involved in these processes are called unit operations. Food processes are associated with diversed activities, but if we carefully analyse them, complicated and differing processed can be broken down to number of small unit operations. For example, heating for instances is used in food industries for numerous purposes required by the food material. Milk needs to be heated to pasteurization temperature to kill all the pathogenic microorganisms. Meat products are frozen to keep them protected from harmful external agents. Fruit products require a mild heat treatment called blanching before further processing to remove the foreign matters and microbial contamination at the surface level. In all the above three cases consideration is given to the extent of heating or cooling to accomplish a defined purpose at different conditions. Thus, this physical process qualifies to be called as a unit operation named 'heat transfer'. A unit operation is dependent on certain physical principles. So, a unit operation is described with the help of a set of quantitative relationships in the form of mathematical equations. A process can be monitored, controlled or modified with the help of these equations.

Classification of Unit Operations

Food products go through different transformation stages for different outcome. Milk is dried to form milk powder. The same milk is fermented to make curd/yoghurt. When acid is added to milk, it gives us cottage cheese. In first case, the change is physical where water is removed by heat treatment. In second case, milk undergoes some biochemical changes by the action of starter microorganisms. Conversion of milk to cheese is a chemical process. So, all the unit operations are broadly classified into three categories.

- Physical: Milling, grinding, sieving, mixing, fluidization, sedimentation, flotation, absorption, adsorption, extraction, rectification, evaporation, drying, extrusion etc.

- Chemical: Coagulation, refining etc.
- Biochemical: Fermentation, pasteurization, sterilization, enzymatic peeling etc.

Depending on the transferred properties, foods undergo transformation due to mass, energy or velocity differences. So, the unit operations are also grouped into following categories.

- Fluid flow processes: Fluidization, filtration, sedimentation, liquid transportation etc.
- Heat transfer processes: Evaporation, drying, concentration etc.
- Mass transfer processes: Absorption, distillation, adsorption, extraction etc.
- Thermodynamic processes: Gas liquefaction, refrigeration etc.
- Mechanical processes: Crushing, pulverization, solid transportation, sieving etc.

In any kind of unit operation, the basic theory is the separation of different components or mixing of these components in homogenous or heterogeneous agricultural products or the transfer of one component to other based on some gradient. To achieve this objective different external agent act on the system that differs according to transferred properties.

Besides, some unit operations such as humidification etc. are complicated in nature wherein simultaneous heat and mass transfer take place. They cannot fall in any categories and are called as complementary unit operations.

Fluid Flow

Many raw materials for foods and many finished foods are in the form of fluids. These fluids have to be transported and processed in the factory. Food technologists must be familiar with the principles that govern the flow of fluids, and with the machinery and equipment that is used to handle fluids.

Importance of Fluid Flow in Food Processing

- Study of fluid behaviour.
- Liquid handling in processing plants.
- Selection of pumps.
- Pipeline design.
- Selection of process equipments based on the properties of fluid.

Science of Fluid Flow

Fluid Mechanics is the branch of engineering science that deal with fluid flow. There are three branches of fluid mechanics:

- Fluid Statics: The study of fluid behaviour at rest.

- Fluid kinematics: The study of flow of fluid without considering its causes.
- Fluid dynamics: It involves the study of flow and the forces causing the flow.

Measurement of Pressure

Devices

- Manometers (Piezometer, simple U-tube and differential manometers).
- Mechanical gauges (Bourdon's gauge and diaphragm).

Manometers are the devices that measure the pressure of fluid at a point which is equal to the mass of standing fluid above it. The pressure difference at two points is measured by balancing the column of same liquid or other liquid.

Manometric Fluid: Mercury

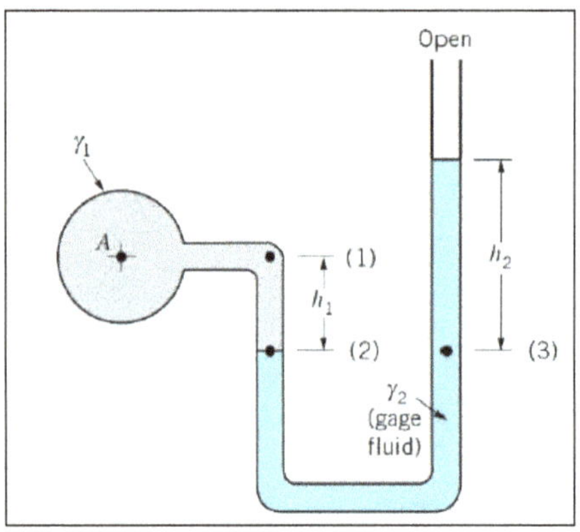

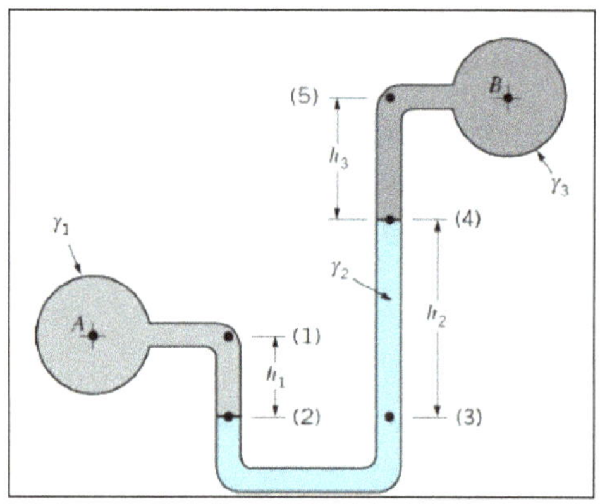

$$p_A + \gamma_1 h_1 - \gamma_2 h_2 = 0$$
$$p_A + \gamma_2 h_2 - \gamma_1 h_1$$

$$p_A + \gamma_1 h_1 - \gamma_2 h_2 - \gamma_3 h_3 = p_B$$
$$p_A - p_B = \gamma_2 h_2 + \gamma_3 h_3 - \gamma_1 h_1$$

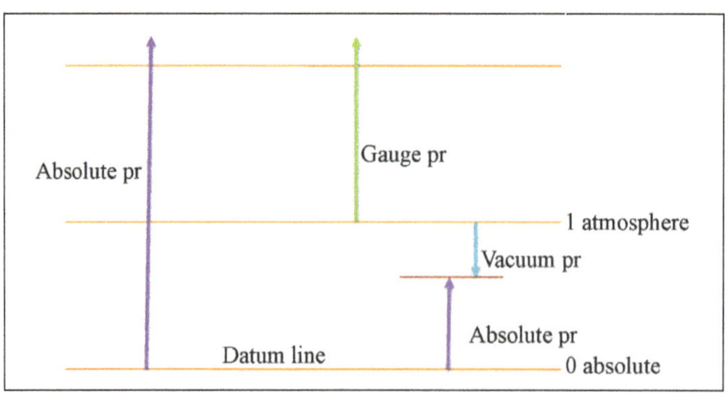

- Absolute pressure: The pressure measured w.r.t. absolute vacuum pressure.
- Gauge pressure: The pressure measured above atmospheric pressure.
- Vacuum pressure: The pressure in short of atmospheric pressure.

$$P_{abs} = P_{atm} + P_{gauge}$$
$$P_{vacuum} = P_{atm} - P_{abs}$$

- Fluid kinematics: A branch of fluid mechanics, which describes the fluid motion and it's consequences without considering the causes of fluid motion.

Flow Patterns

- Stream line: An imaginary line in the flow field whose tangent to the line at any point is the direction of local velocity at that point.
- Path line: Actual path travelled by an individual fluid particle.
- Streak line: The line formed by all fluid particles that have passed through a given fixed point.
- Time line: A set of points form a definite line at any instant in flow field.

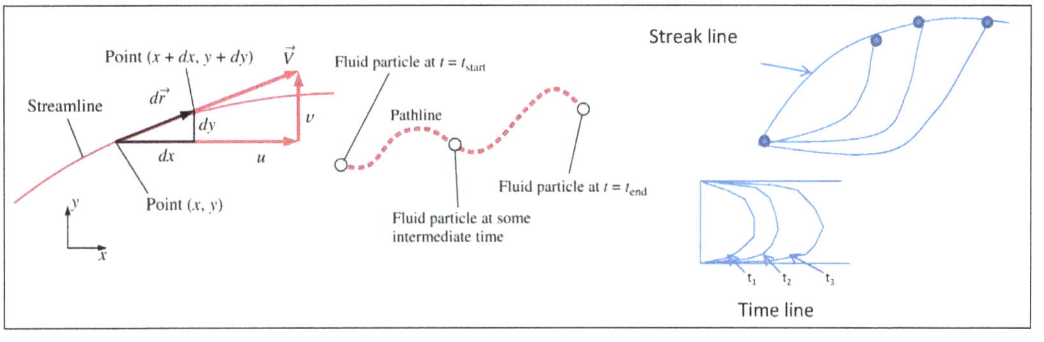

Types of Flow

- Steady state flow: $\left(\frac{\partial V}{\partial t}\right)_{xyz} = 0 \left(\frac{\partial \rho}{\partial t}\right)_{xyz} = 0$
- Un-steady state flow: $\left(\frac{\partial V}{\partial t}\right)_{xyz} \neq 0 \left(\frac{\partial \rho}{\partial t}\right)_{xyz} \neq 0$
- Uniform flow: $\left(\frac{\partial V}{\partial S}\right)_{t} = 0$
- Non-uniform flow: $\left(\frac{\partial V}{\partial S}\right)_{t} \neq 0$
- Compressible flow: density changes with time or space.
- Laminar flow: A set of points flows through a well-defined path or streamline.
- One-two and three dimensional flows:

$$\vec{V} = \vec{u}\hat{i}$$
$$\vec{V} = \vec{u}\hat{i} + \vec{v}\hat{j}$$
$$\vec{V} = \vec{u}\hat{i} + \vec{v}\hat{j} + \vec{w}\hat{k}$$

- Rotational flow: Fluid flows in streamline and it rotates about its own axis.
- Irrational flow: Fluid does not rotate about its own axis.

$$N_{Re} \frac{\text{inertialforce}}{\text{viscousforce}} = \frac{\rho A v^2}{\mu \frac{v}{l} \times A} = \frac{\rho l v}{\mu} \text{ Or, } N_{Re} = \frac{\rho d V}{\mu} \text{ for pipe flow.}$$

Reynold's no. can also be represented in terms of mass flow rate.

$$m = \rho A v \text{ So, } N_{Re} = \frac{4m}{\pi d \mu}$$

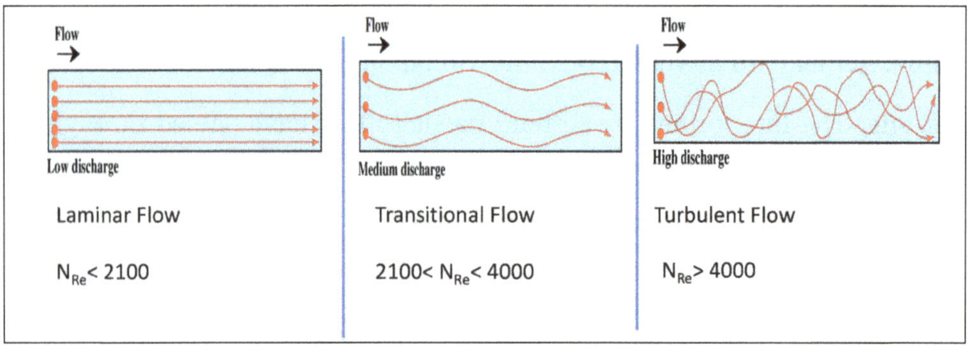

Principle: Conservation of Mass

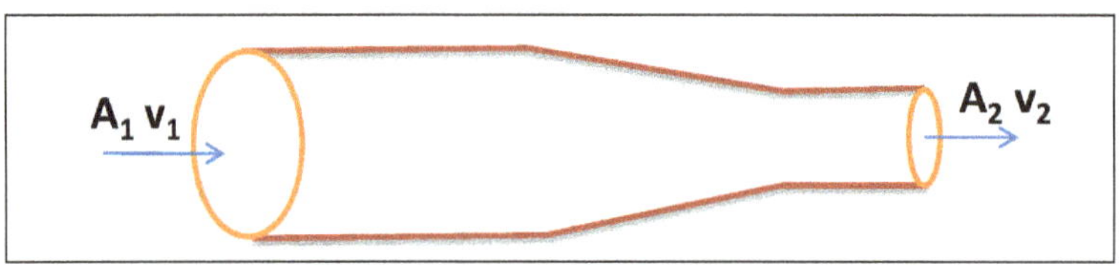

Let us consider two points 1-1 and 2-2 in a pipe line of different cross section:

- Mass flow rate also known as discharge.

  Mass flow rate at 1-1 $m_1 = \rho_1 A_1 V_1$

  Mass flow rate at 2-2 $m_2 = \rho_2 A_2 V_2$

- According to the law of conservation of mass:

  $m_1 = m_2$

  $\therefore \rho_1 A_1 V_1 = \rho_2 A_2 V_2$

- For incompressible fluid.

  $\rho_1 = \rho_2$

So, $A_1 V_1 = A_2 V_2 \quad V \propto \frac{1}{A}$.

The product of cross sectional area and velocity is called discharge,

$$q_1 = q_2$$

Bernoulli's Theorem

Bernoulli's theorem works with the principle of conservation of energy.

$$h + \frac{P}{\rho g} + \frac{v^2}{2g} = \text{Constant}$$

Elevation head (h), Pressure head ($\frac{P}{\rho g}$), Energy head ($\frac{v^2}{2g}$)

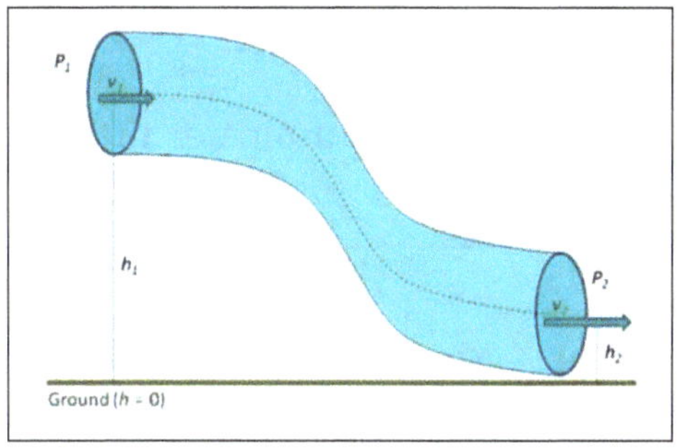

Assumptions:

1) Fluid is ideal and incompressible.

2) Fluid flows in laminar region.

3) Steady state flow.

4) Flow is stream line.

$$h_1 + \frac{P_1}{\rho g} + \frac{v_1^2}{2g} = h_2 + \frac{P_2}{\rho g} + \frac{v_2^2}{2g}$$

Milk is flowing at a rate of 1500 l/h through a pipe of diameter 10 cm at a pressure of 80 kPa. The pipe is reduced to 7 cm diameter at the other end. Neglecting friction loss in pipe and viscosity of liquid, find out the pressure built up at the end of the pipe. Assume milk density as 1030 kg/m³.

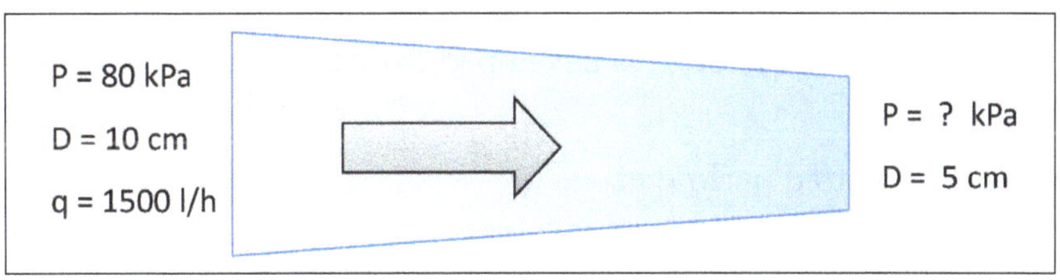

$$q = 1500 l/h = \frac{1500 \times 10^{-3}}{3600} m^3/s = 4.17 \times 10^{-4} m^3/s$$

$$A_1 = \frac{\pi}{4}(0.1)^2 = 7.85 \times 10^{-3} m^2 \qquad A_2 = \frac{\pi}{4}(0.05)^2 = 2.50 \times 10^{-3} m^2$$

$$v_1 = \frac{4.17 \times 10^{-4}}{7.85 \times 10^{-3}} m/s = 0.053 m/s \qquad v_2 = \frac{4.17 \times 10^{-4}}{2.50 \times 10^{-3}} m/s = 0.167 m/s$$

Applying Bernoulli's Equation

$$z_1 + \frac{P_1}{\rho g} + \frac{v_1^2}{2g} z_2 + \frac{P_2}{\rho g} + \frac{v_2^2}{2g} \qquad P_2 = P_1 + \frac{\rho}{2}\left(v_1^2 - v_2^2\right)$$

$$P_2 = 80 + \frac{1030}{2}\left(0.053^2 - 0.167^2\right) = 79.95\,\text{kPa}$$

Energy Losses in Pipes

When fluid moves through a pipe or fittings, it encounters some resistance due to which some energy is lost.

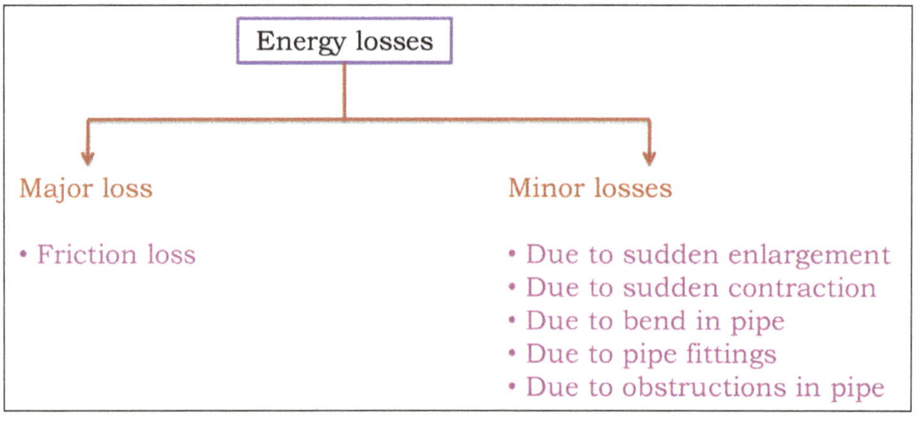

Assuming horizontal pipe and a constant velocity flow,

$$\cancel{h}_1 + \frac{P_1}{\rho g} + \frac{v_1^2}{2g} = \cancel{h}_2 + \frac{P_2}{\rho g} + \frac{v_2^2}{2g} + h_f$$

$$\text{So, } h_f = \frac{P_1 - P_2}{\rho g},$$

A friction factor (Darcy's friction factor, f) is developed to describe the frictional losses in pipe.

$$f = \frac{\text{Drag force per wetted perimeter}}{\rho \times \dfrac{v^2}{2}} = \frac{-\dfrac{dP}{dx}d}{\rho \times \dfrac{v^2}{2}}$$

$$\Rightarrow -\frac{dP}{dx} = \frac{f\rho v^2}{2d}$$

$$\Rightarrow P_1 - P_2 = \frac{fl\rho v^2}{2d}$$

Combining equation $h_f = \frac{P_1 - P_2}{\rho g}$ and $\Rightarrow P_1 - P_2 = \frac{fl\rho v^2}{2d}$ we get, $h_f = f\frac{l}{d}\frac{v^2}{2g}$

Darcy-Weishbach equation can also represented as,

$$h_f = 4f'\frac{l}{d}\frac{v^2}{2g}$$

Where d in this case is diameter of pipe.

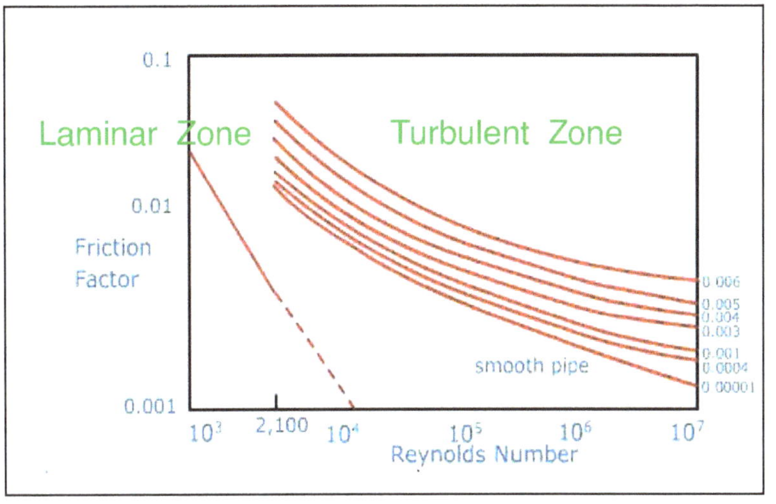

In laminar region ($N_{Re} < 2100$),

$$f' = \frac{16}{N_{Re}}$$

For $4000 < N_{Re} < 10^4$,

$$f' = \frac{0.079}{(N_{Re})^{0.25}}$$

From the Moody's diagram it is seen that Friction factor is a function of Reynold's no. and roughness of pipe.

- Loss due to sudden enlargement: $h_e = \frac{(v_1 - v_2)^2}{2g}$.
- Loss due to sudden contraction: $h_c = k\frac{{v_2}^2}{2g}$ where k, $= \left(\frac{1}{C_c} - 1\right)^2$.
- Loss due to bend in pipe: $h_b = k\frac{v^2}{2g}$.
- Loss due to fittings in pipe: $h_{ft} = k\frac{v^2}{2g}$.
- Loss at the entrance of pipe: $h_i = 0.5\frac{v^2}{2g}$.
- Loss at the exit of pipe: $h_o = \frac{v^2}{2g}$.

Flow Measurements

$$q_{act} = C_d \frac{a_1 a_2}{\sqrt{a_1^2 - a_2^2}} \sqrt{2gh}$$

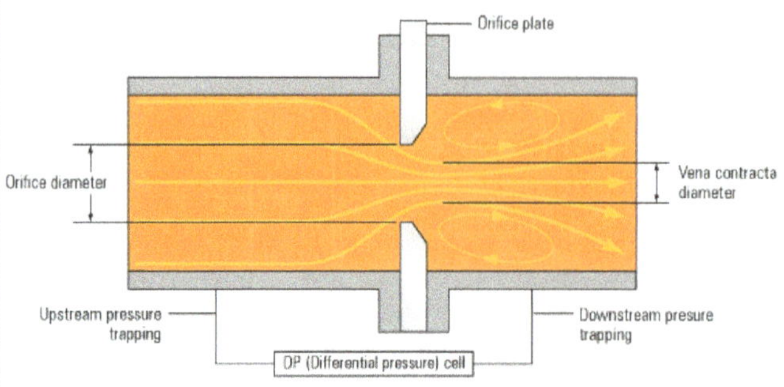

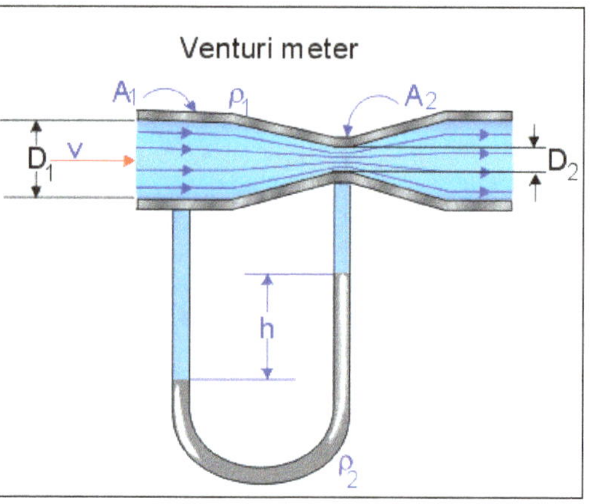

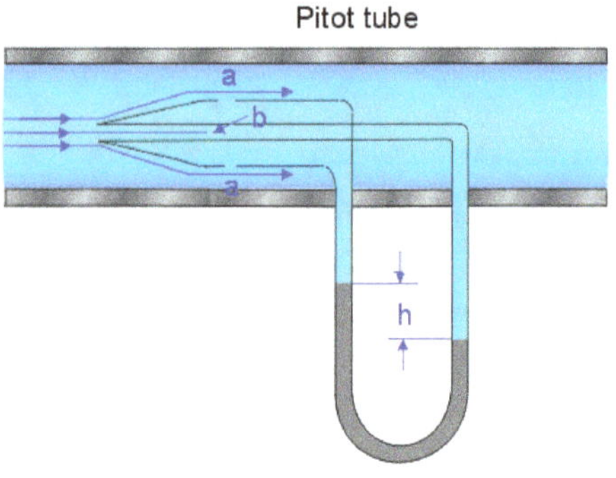

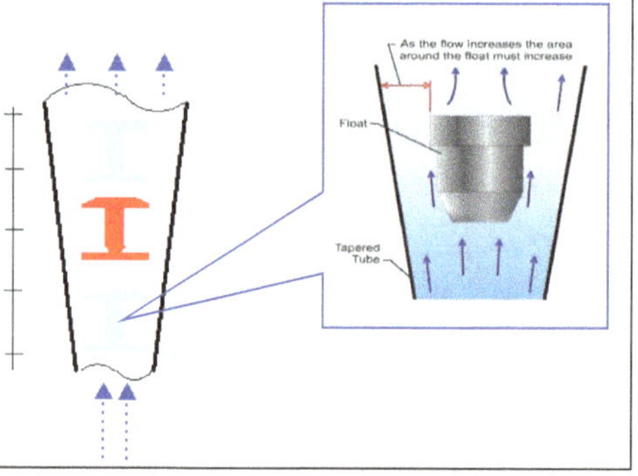

Heat Transfer

Heat transfer is an exchange of thermal energy between two objects. The rate of heat transfer depends upon the temperatures of each entity and the medium through which the thermal energy is being transferred. In cooking, heat transfer refers to heating your food items through a cooking appliance, such as a stove, fryer, microwave, or oven.

Heat Transfer Used in Cooking

Heat transfer is a very important aspect of the cooking process. Heating food destroys potentially harmful bacteria and other microorganisms, which makes food safe to eat and easier to digest. When food or liquids become hot, their molecules absorb energy, begin vibrating rapidly, and start to bounce off of each other. As they collide, heat energy is produced and transferred, which warms and cooks our food.

Types of Heat Transfer

- Conduction
- Convection
- Radiation

Each of these three methods of heat transfer features its own unique characteristics, but there is some crossover between the different types.

Conduction

Conduction is the process of heat being transferred between objects through direct contact, and it's the most common type of heat transfer. For example, in cooking the burners on stoves will conduct heat energy to the bottom of a pan sitting on top of it. From there, the pan conducts heat to its contents.

A deep fryer also uses conduction heating as the hot oil cooks the food when it comes into direct contact with it. Additionally, conduction heat is responsible for moving heat from the outside of the food to the inside. As a result, conduction heat also happens when cooking with convection and radiation heating methods.

Conduction is the slowest method of heat transfer, but the direct contact between the cooking surface and the item to be heated allows food to be cooked from the outside in. When cooking a steak in a cast iron skillet, for example, conduction produces an evenly cooked exterior and a moist, juicy interior that guests are sure to love.

Examples of Conduction Cooking

Here are a few Examples conduction heating:

- Burning your hand on a hot piece of metal.
- Grilling steak, chicken breasts, or pork chops.
- Using ice water to blanch vegetables after steaming to keep them from losing their color.

Convection

Convection combines conduction heat transfer and circulation to force molecules in the air to move from warmer areas to cooler ones. As the molecules closest to the heat source become warm, they rise and are replaced by cooler molecules. There are two types of convection that are based on the movement of the heated molecules.

Natural Convection

Natural convection occurs when molecules at the bottom of a cooking vessel rise and warm while cooler and heavier molecules sink. This creates a circulating current that evenly distributes heat throughout the substance being prepared.

For example, when a pot of water is placed on the stove to boil, conduction heat warms up the pot, which then heats the water molecules inside. As these molecules heat, convection causes them to move away from the interior of the pot as they are replaced by cooler molecules. This continuous current creates convection heat transfer within the water.

Mechanical Convection

Mechanical convection occurs when outside forces circulate heat, which shortens cooking times and cooks food more evenly. Examples of this include stirring liquid in a pot or when a convection oven uses a fan and exhaust system to blow hot air over and around the food before venting it back out.

Examples of Convection Cooking

Here are a few Examples of how heat transfer via convection works:

- Water coming to a boil and circulating in the pot.
- Running cold water over frozen food, which transfers heat into the food to thaw it more quickly.
- Room temperature air moving around frozen food to thaw it.

Radiation Cooking

In cooking, radiation is the process where heat and light waves strike and penetrate your food. As such, there is no direct contact between the heat source and the cooking food. There are two main radiant heat cooking methods: infrared and microwave radiation.

Infrared Radiation

Infrared radiation utilizes an electric or ceramic heating element that gives off electromagnetic energy waves. These waves travel in any direction at the speed of light to quickly heat food, and are mainly absorbed at the surface of whatever you're preparing. Examples of things that create infrared radiation are glowing coals in a fire, toaster ovens, and broilers.

Microwave Radiation

Microwave radiation utilizes short, high-frequency waves that penetrate food, which agitates its water molecules to create friction and transfer heat. If you're heating a solid substance, this heat energy is transferred throughout the food through conduction, while liquids do so through convection.

Microwave heat transfer usually cooks food faster than infrared radiation, as it is able to penetrate foods several inches deep. Keep in mind that microwave radiation works best when cooking small batches of food.

Examples of Radiation Cooking

Here are a few Examples of how heat transfer via radiation works:

- Warming your hands over a fire.
- Lying in the sun to get warm.
- Heating up dinner in the microwave.

The principles of heat transfer are widely used in food processing in many items of equipment. It seems appropriate to discuss these under the various applications that are commonly encountered in nearly every food factory.

Heat Exchangers

In a heat exchanger, heat energy is transferred from one body or fluid stream to another. In the design of heat exchange equipment, heat transfer equations are applied to calculate this transfer of energy so as to carry it out efficiently and under controlled conditions. The equipment goes under many names, such as boilers, pasteurizers, jacketed pans, freezers, air heaters, cookers, ovens and so on. The range is too great to list completely. Heat exchangers are found widely scattered throughout the food process industry.

Continuous-flow Heat Exchangers

It is very often convenient to use heat exchangers in which one or both of the materials that are exchanging heat are fluids, flowing continuously through the equipment and acquiring or giving up heat in passing. One of the fluids is usually passed through pipes or tubes, and the other fluid stream is passed round or across these. At any point in the equipment, the local temperature differences and the heat transfer coefficients control the rate of heat exchange.

The fluids can flow in the same direction through the equipment, this is called parallel flow; they can flow in opposite directions, called counter flow; they can flow at right angles to each other, called cross flow. Various combinations of these directions of flow can occur in different parts of the exchanger. Most actual heat exchangers of this type have a mixed flow pattern, but it is often possible to treat them from the point of view of the predominant flow pattern. Examples of these heat exchangers are illustrated in figure.

In parallel flow, at the entry to the heat exchanger, there is the maximum temperature difference between the coldest and the hottest stream, but at the exit the two streams can only approach each other's temperature. In a counter flow exchanger, leaving streams can approach the temperatures of the entering stream of the other component and so counter flow exchangers are often preferred.

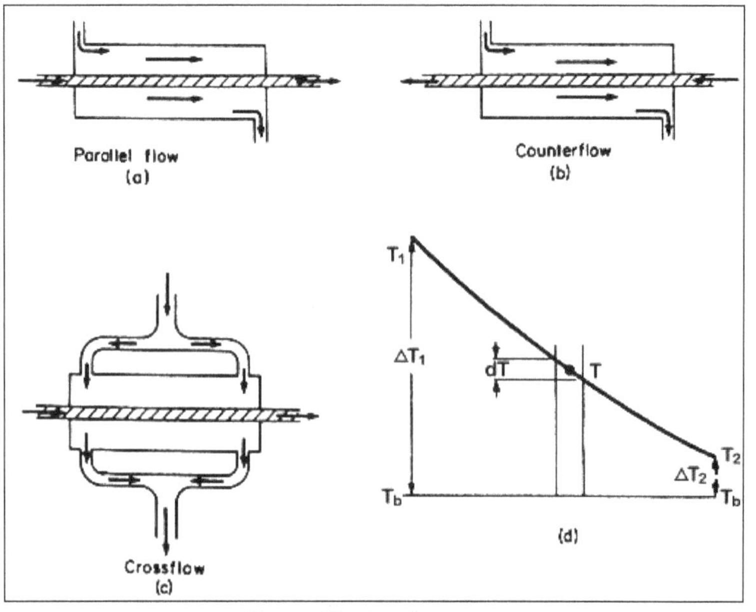

Figure: Heat exchangers.

Applying the basic overall heat transfer equation for the heat transfer in the heat exchanger:

$$q = UA\,\Delta T.$$

Uncertainty at once arises as to the value to be chosen for ΔT, even knowing the temperatures in the entering and leaving streams.

Consider a heat exchanger in which one fluid is effectively at a constant temperature, T_b as illustrated in figure, where time t is on the x-axis and temperature T on the y-axis. Constant temperature in one component can result either from a very high flow rate of this component compared with the other component, or from the component being a vapour such as steam or ammonia condensing at a high rate, or from a boiling liquid. The heat transfer coefficients are assumed to be independent of temperature.

The rate of mass flow of the fluid that is changing temperature is G kgs^{-1}, its specific heat is c_p J $kg^{-1}°C^{-1}$. Over a small length of path of area dA, the mean temperature of the fluid is T and the temperature drop is dT over time dt. The constant temperature fluid has a temperature T_b. The overall heat transfer coefficient is U J $m^{-2}\,s^{-1}\,°C^{-1}$.

Therefore the heat balance over the short length is:

$$c_p\,GdT = U(T - T_b)dA$$

Therefore $(U/c_pG)\,dA = dT/(T - T_b)$

If this is integrated over the length of the tube in which the area changes from A = 0 to A = A, and T changes from T_1 to T_2, we have:

$$(U/c_pG)\,A = \ln[(T_1 - T_b)/(T_2 - T_b)]$$

where ln = $\log_e = \ln(\Delta T_1 / \Delta T_2)$

In which $\Delta T_1 = (T_1 - T_b)$ and $\Delta T_2 = (T_2 - T_b)$

Therefore $c_p G = UA / \ln(\Delta T_1 / \Delta T_2)$

From the overall equation, the total heat transferred per unit time is given by,

$$q = UA\Delta T_m$$

where ΔT_m is the mean temperature difference.

But the total heat transferred per unit is also:

$$q = c_p G(T_1 - T_2)$$

So $q = UA\Delta T_m = c_p G(T_1 - T_2)$

$$= [UA / \ln(\Delta T_1 / \Delta T_2)] \text{ x } (T_1 - T_2)$$

but $(T_1 - T_2)$ can be written $(T_1 - T_b) - (T_2 - T_b)$

so $(T_1 - T_2) = (\Delta T_1 - \Delta T_2)$

Therefore $UA\Delta T_m = UA(\Delta T_1 - \Delta T_2) / \ln(\Delta T_1 / \Delta T_2)$

so that,

$$\Delta T_m = (\Delta T_1 - \Delta T_2) / \ln(\Delta T_1 / \Delta T_2)$$

where ΔT_m is called the log mean temperature difference.

In other words, the rate of heat transfer can be calculated using the heat transfer coefficient, the total area, and the log mean temperature difference. This same result can be shown to hold for parallel flow and counter flow heat exchangers in which both fluids change their temperatures.

The analysis of cross-flow heat exchangers is not so simple, but for these also the use of the log mean temperature difference gives a good approximation to the actual conditions if one stream does not change very much in temperature.

Example: Cooling of milk in a pipe heat exchanger.

Milk is flowing into a pipe cooler and passes through a tube of 2.5cm internal diameter at a rate of 0.4 kgs^{-1}. Its initial temperature is 49°C and it is wished to cool it to 18°C using a stirred bath of constant 10°C water round the pipe. What length of pipe would be required? Assume an overall coefficient of heat transfer from the bath to the milk of 900$Jm^{-2}\ s^{-1}\ °C^{-1}$, and that the specific heat of milk is 3890 $Jkg^{-1}\ °C^{-1}$.

Now,

$$\begin{aligned} q &= c_p G \left(T_1 - T_2\right) \\ &= 3890 \times 0.4 \times \left(49-18\right) \\ &= 48{,}240 Js^{-1} \end{aligned}$$

also, $q = UA\ \Delta T_m$

$$\begin{aligned} \Delta T_m &= \left[(49-10)-(18-10)\right] / \ln\left[(49-10)1(18-10)\right] \\ &= 19.6°C. \end{aligned}$$

therefore, $48{,}240 = 900 \times A \times 19.6$

$$A = 2.73 m^2$$

but $A = \pi DL$

where L is the length of pipe of diameter D.

Now D = 0.025m.

$$L = 2.73/(\pi \times 0.025) = 34.8m.$$

This can be extended to the situation where there are two fluids flowing, one the cooled fluid and the other the heated fluid. Working from the mass flow rates (kgs^{-1}) and the specific heats of the two fluids, the terminal temperatures can normally be calculated and these can then be used to determine ΔT_m and so, from the heat transfer coefficients, the necessary heat-transfer surface.

Example: Water chilling in a counter flow heat exchanger.

In a counter flow heat exchanger, water is being chilled by sodium chloride brine. If the rate of flow of the brine is 1.8 kgs^{-1} and that of the water is 1.05 kgs^{-1}, estimate the temperature to which the water is cooled if the brine enters at –8 °C and leaves at 10C, and if the water enters the exchanger at 32°C.

If the area of the heat-transfer surface of this exchanger is 55 m^2, what is the overall heat transfer coefficient? Take the specific heats to be 3.38 and 4.18$kJkg^{-1}°C^{-1}$ for the brine and the water respectively.

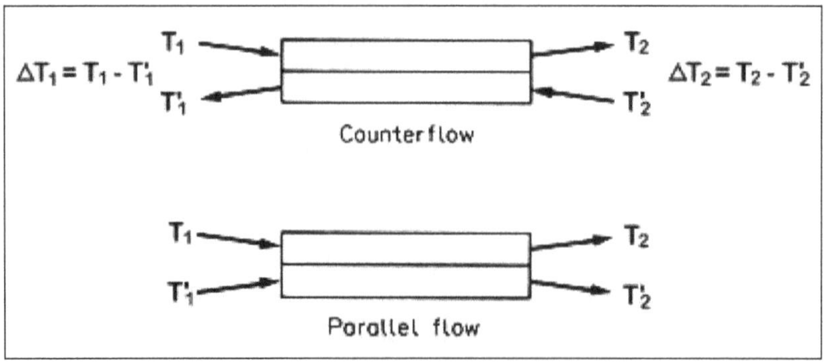

Figure: Diagrammatic heat exchanger.

With heat exchangers a small sketch is often helpful. In this counter flow exchanger, Figure, the brine flows along the top (temperatures T_1 and T_2), and water along the bottom (temperatures T_1' and T_2').

Three temperatures are known and the fourth T_2' can be found from the heat balance:

Brine: entering $T_1 = -8°C$, leaving $T_2 = 10°C$ Water: entering $T_2' = 32°C$, leaving T_1'.

Heat gain in brine = heat loss in water,

$$c_p G\,(T_1 - T_2) = c_p G\,(T_1' - T_2')$$

$$3.38 \times 1.8\,[10 - (-8)] = 4.18 \times 1.05 \times (32 - T_2')$$

$$(32 - T_1') = 25.$$

Therefore T_1' = 7°C.

And so for counter flow,

$$(T_1 - T_1') = \Delta T_1 \; (T_2 - T_2') = \Delta T_2$$

$$\Delta T_1 = [-8\ -7] = 15°C \text{ and } \Delta T_2 = [10 - 32] = 22°C.$$

Therefore, $\Delta T_m = (22 - 15)/\ln(22/15)$

$$= 7/0.382$$

$$= 18.3°C.$$

For the heat exchanger, from the heat balance, heat loss from brine = heat gain to water,

q = heat passed across heat transfer surface $= UA\Delta T_m$

Therefore,

$$3.38 \times 1.8 \times 18 = U \times 55 \times 18.3$$

$$U = 0.11 kJm^{-2}\,s^{-1}°C^{-1}$$

Overall heat transfer coefficient = 110 $Jm^{-2}\,s^{-1}°C^{-1}$.

Parallel flow can be worked out similarly making appropriate adjustments.

In some cases, heat exchanger problems cannot be solved so easily; for example, if the heat transfer coefficients have to be calculated from the basic equations of heat transfer which depend on flow rates and temperatures of the fluids, and the temperatures themselves depend on the heat transfer coefficients. The easiest way to proceed then is to make sensible estimates and to go through the calculations. If the final results are coherent, then the estimates were reasonable. If not, then make better estimates, on the basis of the results, and go through a new set of calculations; and if necessary repeat again until consistent results are obtained. For those with multiple heat exchangers to design, computer programmes are available.

Jacketed Pans

In a jacketed pan, the liquid to be heated is contained in a vessel, which may also be provided with an agitator to keep the liquid on the move across the heat-transfer surface, as shown in Figure.

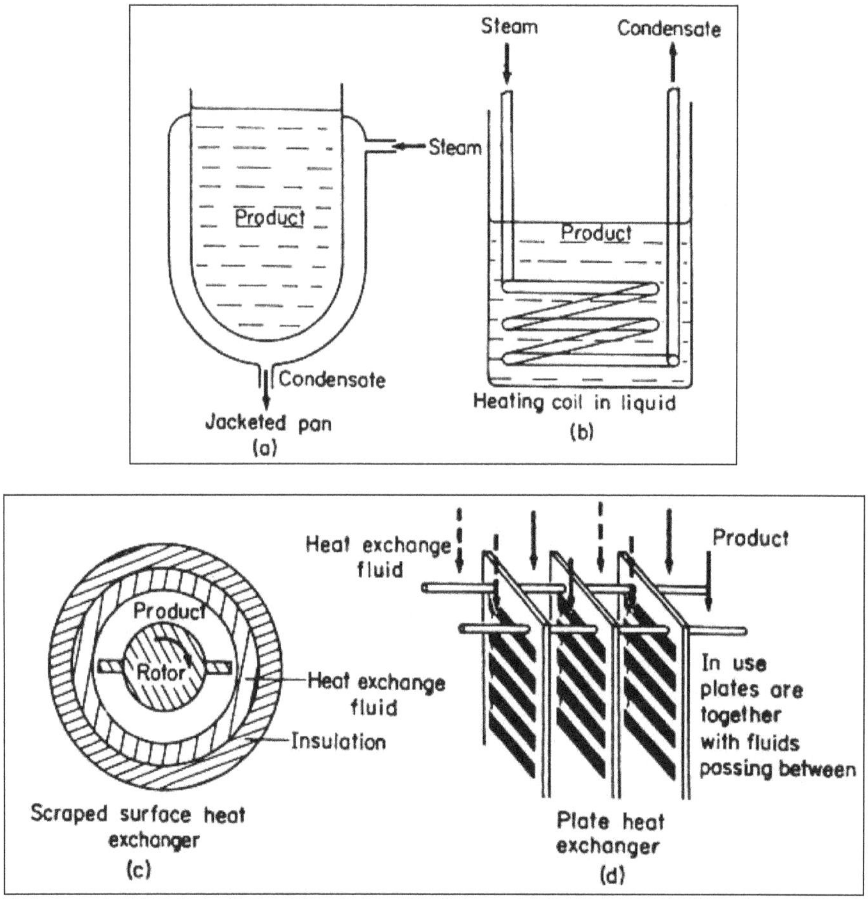

Figure: Heat exchange equipment.

The source of heat is commonly steam condensing in the vessel jacket. Practical considerations are:

- There is the minimum of air with the steam in the jacket.
- The steam is not superheated, as part of the surface must then be used as a de-super heater over which low gas heat transfer coefficients apply rather than high condensing coefficients.
- Steam trapping to remove condensate and air is adequate.

The action of the agitator and its ability to keep the fluid moved across the heat transfer surface are important. Some overall heat transfer coefficients are shown in Table. Save for boiling water, which agitates itself, mechanical agitation is assumed. Where there is no agitation, coefficients may be halved.

Table: Some overall heat transfer coefficients in jacketed pans.

Condensing fluid	Heated fluid	Pan material	Heat transfer coefficients $Jm^{-2}\ s^{-1}{}^{o}C^{-1}$
Steam	Thin liquid	Cast-iron	1800
Steam	Thick liquid	Cast-iron	900
Steam	Paste	Stainless steel	300
Steam	Water, boiling	Copper	1800

Example: Steam required to heat pea soup in jacketed pan.

Estimate the steam requirement as you start to heat 50kg of pea soup in a jacketed pan, if the initial temperature of the soup is 18°C and the steam used is at 100kPa gauge. The pan has a heating surface of 1 m^2 and overall heat transfer coefficient is assumed to be 300 $Jm^{-2}\ s^{-1}{}^{o}C^{-1}$.

From steam tables, saturation temperature of steam at 100kPa gauge = 120°C and latent heat = λ = 2202$kJkg^{-1}$.

$$q = UA\Delta T$$
$$= 300 \times 1 \times (120 - 18)$$
$$= 3.06 \times 10^4\ Js^{-1}$$

Therefore amount of steam,

$$= q/\lambda = (3.06 \times 10^4)/(2.202 \times 10^6)$$

$$= 1.4 \times 10^{-2}\ kgs^{-1}$$

$$= 1.4 \times 10^{-2} \times 3.6 \times 10^3$$

$$= 50kgh^{-1}.$$

This result applies only to the beginning of heating; as the temperature rises less steam will be consumed as ΔT decreases.

The overall heating process can be considered by using the analysis that led up to equation A stirred vessel to which heat enters from a heating surface with a surface heat transfer coefficient which controls the heat flow, follows the same heating or cooling path as does a solid body of high internal heat conductivity with a defined surface heating area and surface heat transfer coefficient.

Example: Time to heat pea soup in a jacketed pan.

In the heating of the pan in Example, estimate the time needed to bring the stirred pea soup up to a temperature of 900C, assuming the specific heat is 3.95 $kJkg^{-1}{}^{o}C^{-1}$.

From equation $(T_2 - T_a)/(T_1 - T_a) = \exp(-h_s At/c\rho V)$:

- T_a = 120°C (temperature of heating medium).
- T_1 = 18°C (initial soup temperature).

- $T_2 = 90°C$ (soup temperature at end of time t).
- $h_s = 300\ Jm^{-2}\ s^{-1}°C^{-1}$.
- $A = 1m^2$, $c = 3.95\ kJkg^{-1}°C^{-1}$.
- $\rho V = 50kg$.

Therefore, $t = \frac{-3.95 \times 10^3 \times 50}{300 \times 1} \times \ln \frac{90-120}{18-120}$

$= (-658) \times (-1.22)\ s$

$= 803s$

$= 13.4min.$

Heating Coils Immersed in Liquids

In some food processes, quick heating is required in the pan, for example, in the boiling of jam. In this case, a helical coil may be fitted inside the pan and steam admitted to the coil as shown in Figure. This can give greater heat transfer rates than jacketed pans, because there can be a greater heat transfer surface and also the heat transfer coefficients are higher for coils than for the pan walls. Examples of the overall heat transfer coefficient U are quoted as:

- 300-1400 for sugar and molasses solutions heated with steam using a copper coil.
- 1800 for milk in a coil heated with water outside.
- 3600 for a boiling aqueous solution heated with steam in the coil.

with the units in these coefficients being $Jm^{-2}\ s^{-1}°C^{-1}$.

Scraped Surface Heat Exchangers

One type of heat exchanger, that finds considerable use in the food processing industry particularly for products of higher viscosity, consists of a jacketed cylinder with an internal cylinder concentric to the first and fitted with scraper blades, as illustrated in Figure. The blades rotate, causing the fluid to flow through the annular space between the cylinders with the outer heat transfer surface constantly scraped. Coefficients of heat transfer vary with speeds of rotation but they are of the order of 900-4000 $Jm^{-2}\ s^{-1}°C^{-1}$. These machines are used in the freezing of ice cream and in the cooling of fats during margarine manufacture.

Plate Heat Exchangers

A popular heat exchanger for fluids of low viscosity, such as milk, is the plate heat exchanger, where heating and cooling fluids flow through alternate tortuous passages between vertical plates as illustrated in Figure. The plates are clamped together, separated by spacing gaskets, and the heating and cooling fluids are arranged so that they flow between alternate plates. Suitable gaskets and channels control the flow and allow parallel or counter current flow in any desired number of passes. A substantial advantage of this type of heat exchanger is that it offers a large transfer surface that is readily accessible for cleaning. The banks of plates are arranged so that they may be taken apart easily. Overall heat transfer coefficients are of the order of 2400-6000 $Jm^{-2} s^{-1}°C^{-1}$.

Thermal Processing

Thermal processing implies the controlled use of heat to increase, or reduce depending on circumstances, the rates of reactions in foods. A common Example is the retorting of canned foods to effect sterilization. The object of sterilization is to destroy all microorganisms, that is, bacteria, yeasts and moulds, in the food material to prevent decomposition of the food, which makes it unattractive or inedible. Also, sterilization prevents any pathogenic (disease-producing) organisms from surviving and being eaten with the food. Pathogenic toxins may be produced during storage of the food if certain organisms are still viable. Microorganisms are destroyed by heat, but the amount of heating required for the killing of different organisms varies. Also, many bacteria can exist in two forms, the vegetative or growing form and the spore or dormant form. The spores are much harder to destroy by heat treatment than are the vegetative forms. Studies of the microorganisms that occur in foods, have led to the selection of certain types of bacteria as indicator organisms. These are the most difficult to kill, in their spore forms, of the types of bacteria which are likely to be troublesome in foods.

A frequently used indicator organism is Clostridium botulinum. This particular organism is a very important food-poisoning organism as it produces a deadly toxin and also its spores are amongst the most heat resistant. Processes for the heat treatment of foodstuffs are therefore examined with respect to the effect they would have on the spores of C. botulinum. If the heat process would not destroy this organism then it is not adequate. As C. botulinum is a very dangerous organism, a selected strain of a non-pathogenic organism of similar heat resistance is often used for testing purposes.

Thermal Death Time

It has been found that microorganisms, including C. botulinum, are destroyed by heat at rates that depend on the temperature, higher temperatures killing spores more quickly. At any given temperature, the spores are killed at different times, some spores being apparently more resistant to heat than other spores. If a graph is drawn, the number of surviving spores against time of holding at any chosen temperature, it is found experimentally that the number of surviving spores fall asymptotically to zero. Methods of handling process kinetics are well developed and if the standard methods are applied to such results, it is found that thermal death of microorganisms follows, for practical purposes, what is called a first-order process at a constant temperature. This implies that the fractional destruction in any fixed time interval is constant. It is thus not possible, in theory at least, to take the time when all of the organisms are actually destroyed. Instead it is practicable, and very useful, to consider the time needed for a particular fraction of the organisms to be killed.

The rates of destruction can in this way be related to:

- The numbers of viable organisms in the initial container or batch of containers.
- The number of viable organisms that can safely be allowed to survive.

Of course the surviving number must be small indeed, very much less than one, to ensure adequate safety. However, this concept, which includes the admissibility of survival numbers of much less than one per container, has been found to be very useful. From such considerations, the ratio of the initial to the final number of surviving organisms becomes the criterion that determines adequate treatment. A combination of historical reasons and extensive practical experience has led to this number being set, for C. botulinum, at 10^{12}:1. For other organisms, and under other circumstances, it may well be different.

The results of experiments to determine the times needed to reduce actual spore counts from 10^{12} to 1 (the lower, open, circles) or to 0 (the upper, closed, circles) are shown in figure.

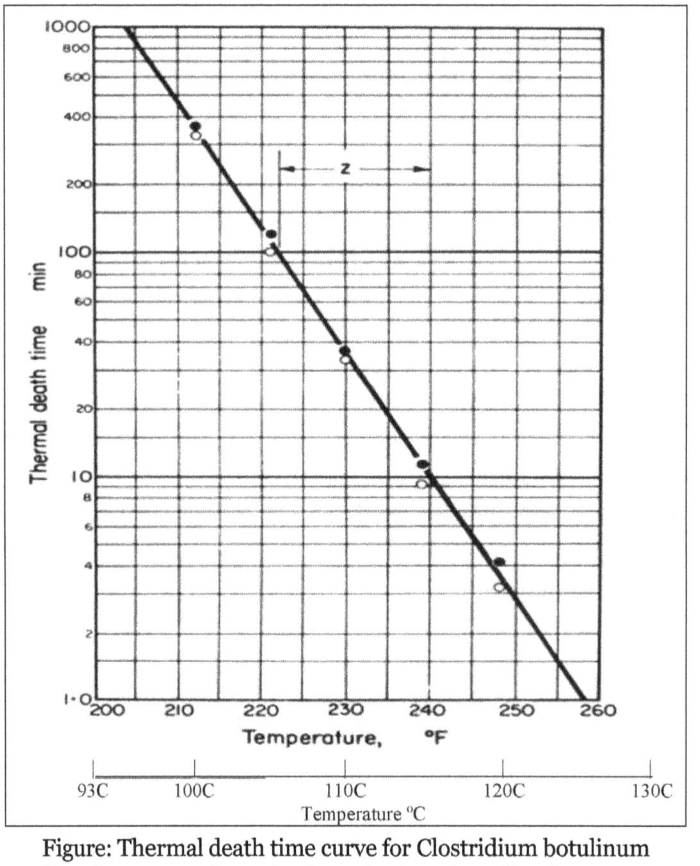

Figure: Thermal death time curve for Clostridium botulinum
Based on research results from the American Can Company.

In this graph, these times are plotted against the different temperatures and it shows that when the logarithms of these times are plotted against temperatures, the resulting graph is a straight line. The mean times on this graph are called thermal death times for the corresponding temperatures. Note that these thermal death times do not represent complete sterilization, but a mathematical concept which can be considered as effective sterilization, which is in fact a survival ratio of 1:10^{12}, and which has been found adequate for safety.

Any canning process must be considered then from the standpoint of effective sterilization. This is done by combining the thermal death time data with the temperature/time relationships at the point in the can that heats slowest. Generally, this point is on the axis of the can and somewhere close to the geometric centre. Using either the unsteady-state heating curves or experimental measurements with a thermocouple at the slowest heating point in a can, the temperature/time graph for the can under the chosen conditions can be plotted. This curve has then to be evaluated in terms of its effectiveness in destroying C. botulinum or any other critical organism, such as thermophilic spore formers, which are important in industry. In this way the engineering data, which provides the temperatures within the container as the process is carried out, are combined with kinetic data to evaluate the effect of processing on the product.

Considering figure, the standard reference temperature is generally selected as 121.1°C (250°F), and the relative time (in minutes) required to sterilize, effectively, any selected organism at 1210C is spoken of as the F value of that organism. In Example, reading from Figure, the F value is about 2.8 min. For any process that is different from a steady holding at 121°C, standard process, the actual attained F values can be worked out by stepwise integration. If the total F value so found is below 2.8 min, then sterilization is not sufficient; if above 2.8 min, the heat treatment is more drastic than it needs to be.

Equivalent Killing Power at other Temperatures

The other factor that must be determined, so that the equivalent killing powers at temperatures different from 121°C can be evaluated, is the dependence of thermal death time on temperature. Experimentally, it has been found that if the logarithm of t, the thermal death time, is plotted against the temperature, a straight-line relationship is obtained. This is shown in Figure and more explicitly in Figure where the x-axis is °C and y-axis is time.

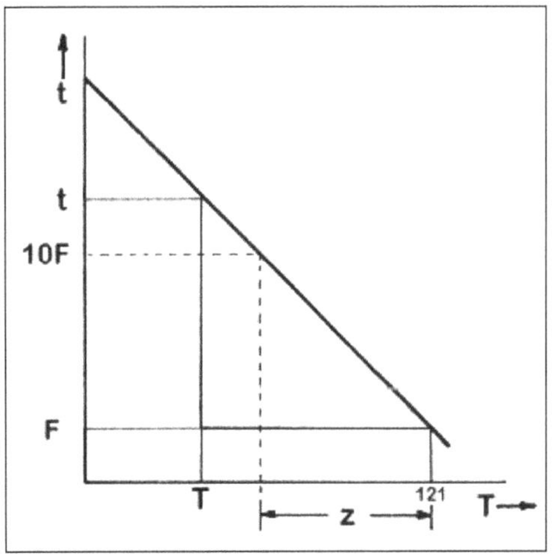

Figure: Thermal death time/temperature relationships.

We can then write from the graph,

$$\log t - \log F = m(121 - T) = \log t / F$$

where t is the thermal death time at temperature T, F is the thermal death time at temperature 121°C (t_{121} = F)and m is the slope of the graph.

Also, if we define the z value as the number of degrees below 121°C at which t increases by a factor of 10, that is by one cycle on a logarithmic graph,

$$t = 10F \text{ when } T = (121 - z)$$

so that, $\log 10F - \log F = \log(10F/F) = 1$

$$1 = m\left[121 - (121 - z)\right]$$

and so z = 1/m

Therefore, $\log(t/F) = (121 - T)/z$

Or,

$$t = F \times 10^{(121-T)/z} = t^{121} \times 10^{(121-T)/z}$$

Now, the fraction of the process towards reaching thermal death, dS, accomplished in time dt is given by $(1/t_1)dt$, where t_1 is the thermal death time at temperature T_1, assuming that the destruction is additive.

That is $dS = (1/t_1)dt$

or $= (1/F)10^{-(121-T)/z}\ dt$

When the thermal death time has been reached, that is when effective sterilization has been achieved,

$$\int dS = 1$$

that is $\int (1/F)10^{-(121-T)/z} dt = 1$,

or, $\int 10^{-(121-T)/z}\ dt = F$.

This implies that the sterilization process is complete, that the necessary fraction of the bacteria/spores have been destroyed, when the integral is equal to F. In this way, the factors F and z can be combined with the time-temperature curve and integrated to evaluate a sterilizing process.

The integral can be evaluated graphically or by stepwise numerical integration. In stepwise numerical integration, the contribution towards F of a period of t min at a temperature T is given by t × $10^{-(121-t)/z}$ Breaking up the temperature-time curve into t_1 min at T_1, t_2 mm at T_2, etc., the total F is given by ,

$$F(\text{i.e.}t_{121}) = t_1 \times 10^{-(121-T_1)/z} + t_2 \times 10^{-(121-T_2)/z} + \ldots\ldots\ldots\ldots$$

This value of F is then compared with the standard value of F for the organism, for example min for C. botulinum in Example, to decide whether the sterilizing procedure is adequate.

Example: Time/Temperature in a can during sterilisation.

In a retort, the temperatures in the slowest heating region of a can of food were measured and were found to be as shown in figure is the retorting adequate, if F for the process is 2.8 mm and z is 10°C?

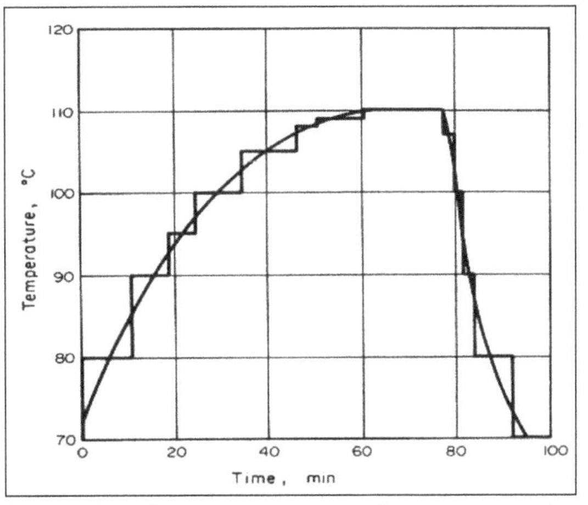

Figure: Time/Temperature curve for can processing.

Approximate stepped temperature increments are drawn on the curve giving the equivalent holding times and temperatures as shown in Table. The corresponding F values are calculated for each temperature step.

Table: Times/Temperatures in Canning Sterilisation.

Temperature	Time				
T (°C)	t (min)	(121 – T)	$10^{-(121-T)/10}$		$t \times 10^{-(121-T)/10}$
80	11	41	7.9×10^{-5}		0.00087
90	8	31	7.9×10^{-4}		0.0063
95	6	26	2.5×10^{-3}		0.015
100	10	21	7.9×10^{-3}		0.079
105	12	16	2.5×10^{-2}		0.30
108	6	13	5.0×10^{-2}		0.30
109	8	12	6.3×10^{-2}		0.50
110	17	11	7.9×10^{-2}		1.34
107	2	14	4.0×10^{-2}		0.08
100	2	21	7.9×10^{-3}		0.016
90	2	31	7.9×10^{-4}		0.0016
80	8	41	7.9×10^{-5}		0.0006
70	6	51	7.9×10^{-6}		0.00005
				Total	2.64

The above results show that the F value for the process = 2.64 so that the retorting time is not quite adequate. This could be corrected by a further 2 min at 110°C (and proceeding as above, this would add $2 \times 10^{-(121-110)/10} = 0.16$, to 2.64, making 2.8).

From the Example, it may be seen that the very sharp decrease of thermal death times with higher temperatures means that holding times at the lower temperatures contribute little to the sterilization. Very long times at temperatures below 90°C would be needed to make any appreciable difference to F, and in fact it can often be the holding time at the highest temperature which virtually determines the F value of the whole process. Calculations can be shortened by neglecting those temperatures that make no significant contribution, although, in each case, both the number of steps taken and also their relative contributions should be checked to ensure accuracy in the overall integration.

It is possible to choose values of F and of z to suit specific requirements and organisms that may be suspected of giving trouble. The choice and specification of these is a whole subject in itself and will not be further discussed. From an engineering viewpoint a specification is set, as indicated above, with an F value and a z value, and then the process conditions are designed to accomplish this.

Pasteurization

Pasteurization is a heat treatment applied to foods, which is less drastic than sterilization, but which is sufficient to inactivate particular disease-producing organisms of importance in a specific foodstuff. Pasteurization inactivates most viable vegetative forms of microorganisms but not heat-resistant spores.

Originally, pasteurization was evolved to inactivate bovine tuberculosis in milk. Numbers of viable organisms are reduced by ratios of the order of 10^{15}:1. As well as the application to inactivate bacteria, pasteurization may be considered in relation to enzymes present in the food, which can be inactivated by heat. The same general relationships as were discussed under sterilization apply to pasteurization. A combination of temperature and time must be used that is sufficient to inactivate the particular species of bacteria or enzyme under consideration. Fortunately, most of the pathogenic organisms, which can be transmitted from food to the person who eats it, are not very resistant to heat.

The most common application of pasteurization is to liquid milk. In the case of milk, the pathogenic organism that is of classical importance is Mycobacterium tuberculosis, and the time/temperature curve for the inactivation of this bacillus is shown in figure.

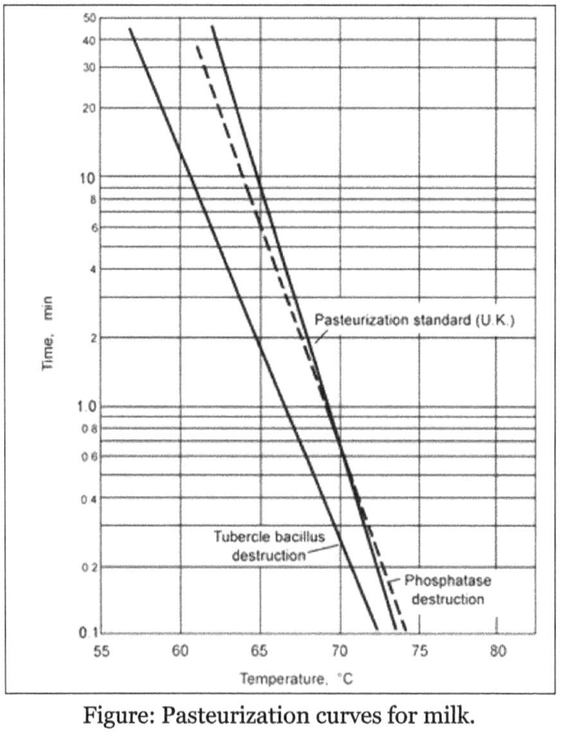

Figure: Pasteurization curves for milk.

This curve can be applied to determine the necessary holding time and temperature in the same way as with the sterilization thermal death curves. However, the times involved are very much shorter, and controlled rapid heating in continuous heat exchangers simplifies the calculations so that only the holding period is really important. For example, 30min at 62.8°C in the older pasteurizing plants and 15sec at 71.7°C in the so-called high temperature/short time (H.T.S.T.) process are sufficient. An even faster process using a temperature of 126.7°C for 4 sec is claimed to be sufficient. The most generally used equipment is the plate heat exchanger and rates of heat transfer to accomplish this pasteurization can be calculated by the methods explained previously.

An enzyme present in milk, phosphatase, is destroyed under somewhat the same temperature/ time conditions as the M. tuberculosis and, since chemical tests for the enzyme can be carried out simply, its presence is used as an indicator of inadequate heat treatment. In this case, the presence or absence of phosphatase is of no significance so far as the storage properties or suitability for human consumption are concerned.

Enzymes are of importance in deterioration processes of fruit juices, fruits and vegetables. If temperature/time relationships, such as those that are shown in Figure for phosphatase, can be determined for these enzymes, heat processes to destroy them can be designed. Most often this is done by steam heating, indirectly for fruit juices and directly for vegetables when the process is known as blanching.

The processes for sterilization and pasteurization illustrate very well the application of heat transfer as a unit operation in food processing. The temperatures and times required are determined and then the heat transfer equipment is designed using the equations developed for heat-transfer operations.

Example: Pasteurisation of milk.

A pasteurization heating process for milk was found, taking measurements and times, to consist essentially of three heating stages being 2min at 64°C, 3min at 650°C and 2min at 66°C. Does this process meet the standard pasteurization requirements for the milk, as indicated in Figure, and if not what adjustment needs to be made to the period of holding at 66°C?

From Figure, pasteurization time's t_T can be read off the UK pasteurisation standard, and from and these and the given times, rates and fractional extents of pasteurization can be calculated.

so At 64°C, t_{64} = 15.7 min:

$$2 \text{ min is } \frac{2}{15.7} = 0.13$$

so At 65°C, t_{65} = 9.2 min:

$$3 \text{ min is } \frac{3}{9.2} = 0.33$$

so At 66°C, t_{66} = 5.4 min:

$$2 \text{ min is } \frac{2}{5.4} = 0.37$$

Total pasteurization extent = (0.13 + 0.33 + 0.37) = 0.83.

Pasteurization remaining to be accomplished = (1 – 0.83) = 0.17.

At 66°C this would be obtained from (0.17 × 5.4) min holding = 0.92 min.

So an additional 0.92 min (or approximately 1 min) at 66°C would be needed to meet the specification.

Refrigeration, Chilling and Freezing

Rates of decay and of deterioration in foodstuffs depend on temperature. At suitable low temperatures, changes in the food can be reduced to economically acceptable levels. The growth and metabolism of microorganisms is slowed down and if the temperature is low enough, growth of all microorganisms virtually ceases. Enzyme activity and chemical reaction rates (of fat oxidation, for example) are also very much reduced at these temperatures. To reach temperatures low enough for deterioration virtually to cease, most of the water in the food must be frozen. The effect of chilling is only to slow down deterioration changes. In studying chilling and freezing, it is necessary to look first at the methods for obtaining low temperatures, i.e. refrigeration systems, and then at the coupling of these to the food products in chilling and freezing.

Refrigeration Cycle

The basis of mechanical refrigeration is the fact that at different pressures the saturation (or the condensing) temperatures of vapours are different as clearly shown on the appropriate vapour-pressure/temperature curve. As the pressure increases condensing temperatures also increase. This fact is applied in a cyclic process, which is illustrated in figure.

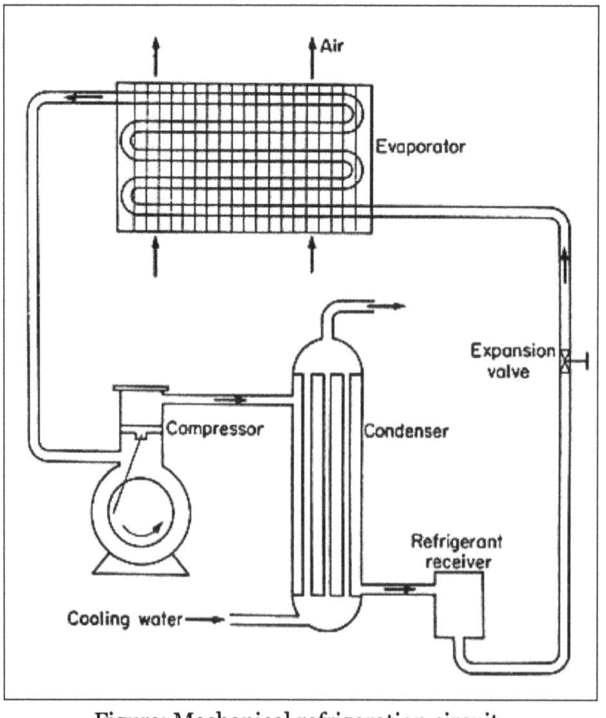

Figure: Mechanical refrigeration circuit.

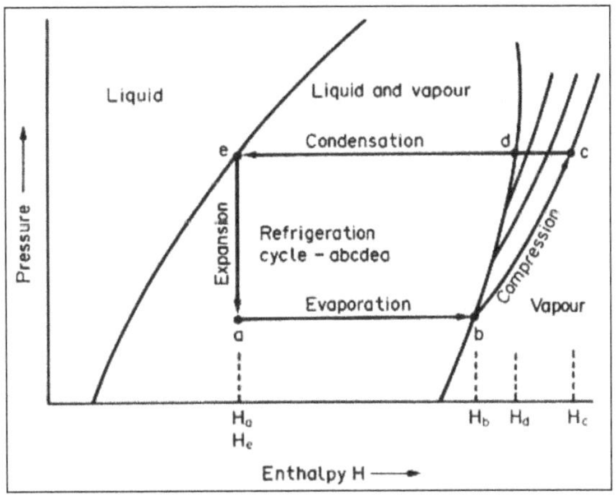

Figure: Pressure/enthalpy chart.

The process can be followed on the pressure/enthalpy (P/H) chart shown in figure. This is a thermodynamic diagram that looks very complicated at first sight but which in fact can make calculations straightforward and simple. For the present purposes the most convenient such chart is the one shown with pressure as the vertical axis (for convenience on a logarithmic scale) and enthalpy on the horizontal axis. On such a diagram, the properties of the particular refrigerant can be plotted, including the interphase equilibrium lines such as the saturated vapour line, which are important as refrigeration depends on evaporation and condensation.

To start with the evaporator; in this the pressure above the refrigerant is low enough so that evaporation of the refrigerant liquid to a gas occurs at some suitable low temperature determined by the requirements of the product. This temperature might be, for example –18°C in which case the corresponding pressures would be for ammonia 229kPa absolute and for tetrafluoroethane (also known as refrigerant 134A) 144kPa absolute. Evaporation then occurs and this extracts the latent heat of vaporization for the refrigerant from the surroundings of the evaporator and it does this at the appropriate low temperature. This process of heat extraction at low temperature represents the useful part of the refrigerator. On the pressure/enthalpy chart this is represented by ab at constant pressure (the evaporation pressure) in which 1 kg of refrigerant takes in $(H_b - H_a)$ kJ. The low pressure necessary for the evaporation at the required temperature is maintained by the suction of the compressor.

The remainder of the process cycle is included merely so that the refrigerant may be returned to the evaporator to continue the cycle. First, the vapour is sucked into a compressor which is essentially a gas pump and which increases its pressure to exhaust it at the higher pressure to the condensers. This is represented by the line bc which follows an adiabatic compression line, a line of constant entropy and work equivalent to $(H_c - H_b)$ kJ kg^{-1} has to be performed on the refrigerant to effect the compression. The higher pressure might be, for example, 1150 kPa absolute for ammonia, or 800 kPa absolute for refrigerant 134A, and it is determined by the temperature at which cooling water or air is available to cool the condensers.

To complete the cycle, the refrigerant must be condensed, giving up its latent heat of vaporization to some cooling medium. This is carried out in a condenser, which is a heat exchanger cooled generally by water or air. Condensation is shown on figure along the horizontal line (at the constant condenser pressure), at first cd cooling the gas and then continuing along de until the refrigerant is completely condensed at point e. The total heat given out in this from refrigerant to condenser water is $(H_c - H_e) = (H_c - H_a)$kJkg^{-1}. The condensing temperature, corresponding to the above high pressures, is about 30oC and so in this Example cooling water at about 20°C, could be used, leaving sufficient temperature difference to accomplish the heat exchange in equipment of economic size.

This process of evaporation at a low pressure and corresponding low temperature, followed by compression, followed by condensation at around atmospheric temperature and corresponding high pressure, is the refrigeration cycle. The high-pressure liquid then passes through a nozzle from the condenser or high-pressure receiver vessel to the evaporator at low pressure, and so the cycle continues. Expansion through the expansion valve nozzle is at constant enthalpy and so it follows the vertical line ea with no enthalpy added to or subtracted from the refrigerant. This line at constant enthalpy from point e explains why the point a is where it is on the pressure line, corresponding to the evaporation (suction) pressure.

By adjusting the high and low pressures, the condensing and evaporating temperatures can be selected as required. The high pressure is determined: by the available cooling-water temperature, by the cost of this cooling water and by the cost of condensing equipment. The evaporating pressure is determined by either the low temperature that is required for the product or by the rate of cooling/freezing that has to be provided. Low evaporating temperatures mean higher power requirements for compression and greater volumes of low-pressure vapours to be handled therefore larger compressors, so that the compression is more expensive. It must also be remembered that, in actual operation, temperature differences must be provided to operate both the evaporator and the condenser. There must be lower pressures than the pressure that corresponds to the evaporating coil temperature in the compressor suction line, and higher pressures in the compressor discharge than the pressures that correspond to the condenser temperature.

Overall, the energy side of the refrigeration cycle can therefore be summed up:

- Heat taken in from surroundings at the (low) evaporator temperature and pressure $(H_b - H_a)$.
- Heat equivalent to the work done by the compressor $(H_c - H_b)$.
- Heat rejected at the (high) compressor pressure and temperature $(H_c - H_e)$.

A useful measure is the ratio of the heat taken in at the evaporator (the useful refrigeration), $(H_b - H_a)$, to the energy put in by the compressor which must be paid for $(H_c – H_b)$. This ratio is called the coefficient of performance (COP). The unit commonly used to measure refrigerating effect is the ton of refrigeration = 3.52 kW. It arises from the quantity of energy to freeze 2000lb of water in one day (2000lb is called 1 short ton).

Example: Freezing of fish.

It is wished to freeze 15 tonnes of fish per day from an initial temperature of 10°C to a final temperature of – 8°C using a stream of cold air. Estimate the maximum capacity of the refrigeration plant required, if it is assumed that the maximum rate of heat extraction from the product is twice the average rate. If the heat transfer coefficient from the air to the evaporator coils, which form the heat exchanger between the air and the boiling refrigerant, is 22 $Jm^{-2} s^{-1} °C^{-1}$, calculate the surface area of evaporator coil required if the logarithmic mean temperature drop across the coil is 12°C.

From the tabulated data the specific heat of fish is 3.18 $kJkg^{-1} °C^{-1}$ above freezing and 1.67 $kJkg^{-1} °C^{-1}$ below freezing, and the latent heat is 276$kJkg^{-1}$.

Enthalpy change in fish = heat loss above freezing + heat loss below freezing + latent heat:

$$= (10 \times 3.18) + (8 \times 1.67) + 276$$

$$= 31.8 + 13.4 + 276$$

$$= 321.2 kJkg^{-1}.$$

Total heat removed in freezing = 15 × 1000 × 321.2 = 4.82 x 10^6 $kJday^{-1}$.

Average rate of heat removal = $(4.82 \times 10^6)/(24 \times 60 \times 60) = 55.8 kJs^{-1}$.

If maximum is twice average,

then maximum = 111.6kJs^{-1} = q.

Coil rate of heat transfer $q = UA\Delta T_m$

and so $A = q / U\Delta T_m$.

$$= (111.6 \times 10^3)/(22 \times 12)$$

$$= 0.42 \times 10^3 m^2$$

$$= 420 m^2.$$

The great advantage of tracing the cycle on the pressure/enthalpy diagram is that from the numerical co-ordinates of the various cycle points, performance parameters can be read off or calculated readily.

Example: Rate of boiling of refrigerant.

Ammonia liquid is boiling in an evaporator under an absolute pressure of 120 kPa. Find the temperature and the volumetric rate of evolution of ammonia gas if the heat extraction rate from the surroundings is 300 watts.

The boiling temperature of ammonia at a pressure of 120kPa is – 30°C, so this is the evaporator temperature.

Also from the chart, the latent heat of evaporation, = (enthalpy of saturated vapour - enthalpy of saturated liquid),

$$= 1.72 - 0.36$$

$$= 1.36 \text{ MJ } kg^{-1}$$

$$= 1.36 \times 10^3 kJkg^{-1}$$

For a heat-removal rate of 300 watts = 0.3 kJs^{-1}

the ammonia evaporation rate = (0.3/1360)

$= 2.2 \times 10^{-4}$ kg s^{-1}.

The chart shows at the saturated vapour point for the cycle that the specific volume of ammonia vapour is:

0.98 m^3 kg^{-1}

and so the volumetric rate of ammonia evolution is:

$$2.2 \times 10^{-4} \times 0.98 = \underline{2.16 \times 10^{-4}\ m^3\ s^{-1}}.$$

So far we have been talking of the theoretical cycle. Real cycles differ by, for example, pressure drops in piping, superheating of vapour to the compressor and non-adiabatic compression, but these are relatively minor. Approximations quite good enough for our purposes can be based on the theoretical cycle. If necessary, allowances can be included for particular inefficiencies.

The refrigerant vapour has to be compressed so it can continue round the cycle and be condensed. From the refrigeration demand, the weight of refrigerant required to be circulated can be calculated; each kg s^{-1} extracts so many Js^{-1} according to the value of (H_b - H_a). From the volume of this refrigerant, the compressor displacement can be calculated, as the compressor has to handle this volume. Because the compressor piston cannot entirely displace all of the working volume of the cylinder, there must be a clearance volume; and because of inefficiencies in valves and ports, the actual amount of refrigerant vapour taken in is less than the theoretical. The ratio of these, actual amount taken in, to the theoretical compressor displacement, is called the volumetric efficiency of the compressor. Both mechanical and volumetric efficiencies can be measured, or taken from manufacturer's data, and they depend on the actual detail of the equipment used.

Consideration of the data from the thermodynamic chart and of the refrigeration cycle enables quite extensive calculations to be made about the operation.

Example: Operation of a compressor in a refrigeration system.

To meet the requirements of Example, calculate the speed at which it would be necessary to run a six-cylinder reciprocating ammonia compressor with each cylinder having a 10cm bore (diameter) and a 12cm stroke (length of piston travel), assuming a volumetric efficiency of 80%. The condensing temperature is 30°C (determined from the available cooling water temperature) and the evaporating temperature needed is – 15°C. Calculate also the theoretical coefficient of performance of this refrigeration system.

From the chart, the heat extracted by ammonia at the evaporating temperature of –15°C is (1.74-0.63),

= 1.11$MJkg^{-1}$

= 1.11 × 103 $kJkg^{-1}$

- Maximum rate of refrigeration = 111.6 kJ s^{-1}.

- Rate of refrigerant circulation = 111.6/1110 = 0.100kgs^{-1}.
- Specific volume of refrigerant (from chart) = 0.49 m^3 kg^{-1}.
- Theoretical displacement volume = 0.49 × 0.100 = 0.049m^3 s^{-1}.
- Actual displacement needed = (0.049 × 100)/80 = 61.3 x 10^{-3}m^3 s^{-1}.
- Volume of cylinders swept out per rev = (π/4) x (0.10)^2 × (0.12 × 6) = 5.7 × 10^{-3}.
- Speed of compressor = (61.3 × 10^{-3})/(5.7 × 10^{-3})

  = 10.75 revs.s^{-1}

  = 645 rev.min^{-1}.

Coefficient of performance = (heat energy extracted in evaporator)/(heat equivalent of theoretical energy input to the compressor),

$$=(H_b - H_a)/(H_c - H_b).$$

Under cycle conditions, from chart, the coefficient of performance,

= (1.74- 0.63)/(1.97 -1.74)

= 1.11/0.23

= 4.8.

Performance Characteristics

Variations in load and in the evaporating or condensing temperatures are often encountered when considering refrigeration systems. Their effects can be predicted by relating them to the basic cycle.

If the heat load increases in the cold store, then the temperature tends to rise and this increases the amount of refrigerant boiling off. If the compressor cannot move this, then the pressure on the suction side of the compressor increases and so the evaporating temperature increases tending to reduce the evaporation rate and correct the situation. However, the effect is to lift the temperature in the cold space and if this is to be prevented additional compressor capacity is required.

As the evaporating pressure, and resultant temperature, change so the volume of vapour per kilogram of refrigerant changes. If the pressure decreases, this volume increases, and therefore the refrigerating effect, which is substantially determined by the rate of circulation of refrigerant, must also decrease. Therefore if a compressor is required to work from a lower suction pressure its capacity is reduced, and conversely. So at high suction pressures giving high circulation rates, the driving motors may become overloaded because of the substantial increase in quantity of refrigerant circulated in unit time.

Changes in the condenser pressure have relatively little effect on the quantity of refrigerant circulated. However, changes in the condenser pressure and also decreases in suction pressure, have quite a substantial effect on the power consumed per ton of refrigeration. Therefore for an economical plant, it is important to keep the suction pressure as high as possible, compatible with the product requirement for low temperature or rapid freezing, and the condenser pressure as low as possible compatible with the available cooling water or air temperature.

Refrigerants

Although in theory a considerable number of fluids might be used in mechanical refrigeration, and historically quite a number including cold air and carbon dioxide have been, those actually in use today are only a very small number. Substantially they include only ammonia, which is used in many large industrial systems, and approved members of a family of halogenated hydrocarbons containing differing proportions of fluorine and chosen according to the particular refrigeration duty required. The reasons for this very small group of practical refrigerants are many. Important reasons are:

- Actual vapour pressure/temperature curve for the refrigerant which determines the pressures between which the system must operate for any particular pair of evaporator and condenser temperatures.
- Refrigerating effect per cubic metre of refrigerant pumped around the system which in turn governs the size of compressors and piping, and the stability and cost of the refrigerant itself.

Ammonia is in most ways the best refrigerant from the mechanical point of view, but its great problem is its toxicity. In working spaces, such as encountered in air conditioning, the halogenated hydrocarbons (often known, after the commercial name, as Freons) are very often used because of safety considerations. They are also used in domestic refrigerators. Environmental problems, in particular due to the effects of any chlorine derivatives and their long lives in the upper atmosphere, have militated heavily, in recent years, against some formerly common halogenated hydrocarbons. Only a very restricted selection has been judged safe.

Mechanical Equipment

Compressors are just basically vapour pumps and much the same types as shown in Figure for liquid pumps are encountered. Their design is highly specialized, particular problems arising from the lower density and viscosity of vapours when compared with liquids. The earliest designs were reciprocating machines with pistons moving horizontally or vertically, at first in large cylinders and at modest speeds, and then increasingly at higher speeds in smaller cylinders.

An important aspect in compressor choice is the compression ratio, being the ratio of the absolute pressure of discharge from the compressor, to the inlet suction pressure. Reciprocating compressors can work effectively at quite high compression ratios (up to 6 or 7 to 1). Higher overall compression ratios are best handled by putting two or more compressors in series and so sharing the overall compression ratio between them.

For smaller compression ratios and for handling the large volumes of vapours encountered at low temperatures and pressures, rotary vane compressors are often used, and for even larger volumes, centrifugal compressors, often with many stages, can be used. A recent popular development is the screw compressor, analogous to a gear pump, which has considerable flexibility. Small systems are often "hermetic", implying that the motor and compressor are sealed into one casing with the refrigerant circulating through both. This avoids rotating seals through which refrigerant can leak. The familiar and very dependable units in household refrigerators are almost universally of this type.

Refrigeration Evaporator

The evaporator is the only part of the refrigeration equipment that enters directly into food processing operations. Heat passes from the food to the heat transfer medium, which may be air or liquid, thence to the evaporator and so to the refrigerant. Thermal coupling is in some cases direct, as in a plate freezer. In this, the food to be frozen is placed directly on or between plates, within which the refrigerant circulates. Another familiar Example of direct thermal coupling is the chilled slab in a shop display.

Generally, however, the heat transfer medium is air, which moves either by forced or natural circulation between the heat source, the food and the walls warmed by outside air, and the heat sink which is the evaporator. Sometimes the medium is liquid such as in the case of immersion freezing in propylene glycol or in alcohol/water mixtures. Then there are some cases in which the refrigerant is, in effect, the medium such as immersion or spray with liquid nitrogen. Sometimes there is also a further intermediate heat transfer medium, so as to provide better control, or convenience, or safety. An Example is in some milk chillers, where the basic refrigerant is ammonia; this cools glycol, which is pumped through a heat exchanger where it cools the milk. Some types of evaporator systems are in figure.

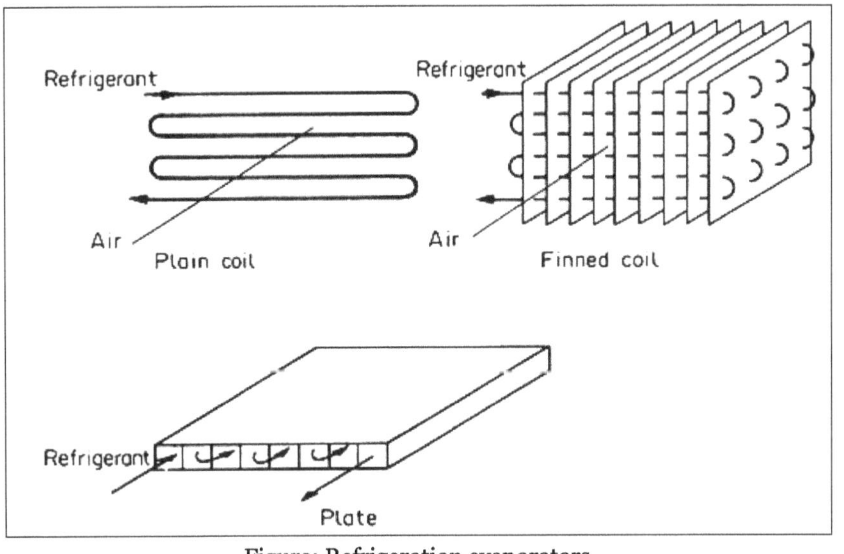

Figure: Refrigeration evaporators.

In the freezer or chiller, the heat transfer rates both from the food and to the evaporator, depend upon fluid or gas velocities and upon temperature differences. The values for the respective heat transfer coefficients can be estimated by use of the standard heat transfer relationships. Typical values that occur in freezing equipment are given in table.

Table: Expected Surface Heat Transfer Coefficients (H_s).

	J m-2 s^{-1} $^{o}C^{-1}$
Still-air freezing (including radiation to coils)	9
Still-air freezing (no radiation)	6

Air-blast freezing 3 ms^{-1}	18
Air-blast freezing 5 ms^{-1}	30
Liquid-immersion freezing	600
Plate freezing	120

Temperature differences across evaporators are generally of the order 3 – 10°C. The evaporator surfaces are often extended by the use of metal fins that are bonded to the evaporator pipe surface. The reason for this construction is that the relatively high metal conductance, compared with the much lower surface conductance from the metal surface to the air, maintains the fin surface substantially at the coil temperature. A slight rise in temperature along the fin can be accounted for by including in calculations a fin efficiency factor. The effective evaporator area is then calculated by the relationship,

$$A = A_p + \phi A_s$$

where A is the equivalent total evaporator surface area, A_p is the coil surface area, called the primary surface, As is the fin surface area, called the secondary surface and ϕ (phi) is the fin efficiency. Values of ϕ lie between 65% and 95% in the usual designs.

Chilling

Chilling of foods is a process by which their temperature is reduced to the desired holding temperature just above the freezing point of food, usually in the region of -2 to 2°C. Many commercial chillers operate at higher temperatures, up to 10-12°C. The effect of chilling is only to slow down deterioration changes and the reactions are temperature dependent. So the time and temperature of holding the chilled food determine the storage life of the food.

Rates of chilling are governed by the laws of heat transfer,. It is an Example of unsteady-state heat transfer by convection to the surface of the food and by conduction within the food itself. The medium of heat exchange is generally air, which extracts heat from the food and then gives it up to refrigerant in the evaporator. Rates of convection heat transfer from the surface of food and to the evaporator are much greater if the air is in movement, being roughly proportional to $v^{0.8}$.

To calculate chilling rates it is therefore necessary to evaluate:

- Surface heat transfer coefficient.
- Resistance offered to heat flow by any packaging material that may be placed round the food.
- Appropriate unsteady-state heat conduction equation.

Although the shapes of most foodstuffs are not regular, they often approximate the shapes of slabs, bricks, spheres and cylinders.

Example: Chilling of fresh apples.

Before apples are loaded into a cool store, it is wished to chill them to a central temperature of 5°C so as to avoid problems of putting warm apples with the colder ones in storage. The apples, initially at 25°C, are considered to be spheres of 7cm diameter and the chilling is to be carried out using air at –1 °C and at a velocity that provides a surface heat-transfer coefficient of 30Jm^{-2} s^{-1} $°C^{-1}$.

The physical properties of the apples: k = 0.5J m^{-1} s^{-1} °C^{-1}, ρ = 930kgm^{-3}, c = 3.6 kJ kg^{-1}°C^{-1}. Calculate the time necessary to chill the apples so that their centres reach 5°C.

This is an Example of unsteady-state cooling and can be solved by application of figure,

$$Bi = \frac{h_s r}{k} = \frac{30 \times 0.035}{0.5} = 2.1$$

1/Bi = 0.48,

$$(T - T_o)/(T_1 - T_o) = [5-(-1)]/[25-(-1)] = 0.23$$

and so reading from Figure, the Fourier Number,

$$Fo = 0.46 = kt / rcr^2$$
$$t = Forcr^2 / k$$

so t = [0.46 × 3600 ×930 × (0.035)2]/0.5

$$= 3773s$$
$$= \underline{1.05h}.$$

A full analysis of chilling must, in addition to heat transfer, take mass transfer into account if the food surfaces are moist and the air is unsaturated. This is a common situation and complicates chilling analysis.

Freezing

Water makes up a substantial proportion of almost all foodstuffs and so freezing has a marked physical effect on the food. Because of the presence of substances dissolved in the water, food does not freeze at one temperature but rather over a range of temperatures. At temperatures just below the freezing point of water, crystals that are almost pure ice form in the food and so the remaining solutions become more concentrated. Even at low temperatures some water remains unfrozen, in very concentrated solutions.

In the freezing process, added to chilling is the removal of the latent heat of freezing. This latent heat has to be removed from any water that is present. Since the latent heat of freezing of water is 335 kJkg^{-1}, this represents the most substantial thermal quantity entering into the process. There may be other latent heats, for example the heats of solidification of fats which may be present, and heats of solution of salts, but these are of smaller magnitude than the latent heat of freezing of water. Also the fats themselves are seldom present in foods in as great a proportion as water.

Because of the latent heat removal requirement, the normal unsteady-state equations cannot be applied to the freezing of foodstuffs. The coefficients of heat transfer can be estimated by the following equation:

$$1/h_s = 1/hc + (x/k) + 1/h_r$$

where h_s is the total surface heat-transfer coefficient, h_c is the convection heat transfer coefficient, x is the thickness of packing material, k is the thermal conductivity of the packing material and h_r is the radiation heat-transfer coefficient.

A full analytical solution of the rate of freezing of food cannot be obtained. However, an approximate solution, due to Plank, is sufficient for many practical purposes. Plank assumed that the freezing process:

- Commences with all of the food unfrozen but at its freezing temperature.
- Occurs sufficiently slowly for heat transfer in the frozen layer to take place under steady-state conditions.

Making these assumptions, freezing rates for bodies of simple shapes can be calculated. As an Example of the method, the time taken to freeze to the centre of a slab whose length and breadth are large compared with the thickness, will be calculated.

Rates of heat transfer are equal from either side of the slab. Assume that at time t a thickness x of the slab of area A has been frozen as shown in figure. The temperature of the freezing medium is T_a. The freezing temperature of the foodstuff is T, and the surface temperature of the food is T_s. The thermal conductivity of the frozen food is k, λ is the latent heat of the foodstuff and ρ is its density.

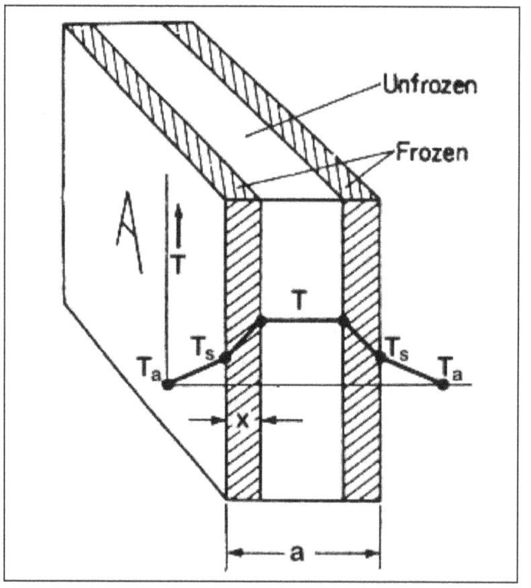

Figure: Freezing of a slab.

The rate of movement of the freezing boundary multiplied by the latent heat equals the rate of heat transfer of heat to the boundary:

$$q = A\, \lambda\rho\, dx / dt$$

Now, all of the heat removed at the freezing boundary must be transmitted to the surface through the frozen layer; if the frozen layer is in the steady-state condition we have:

$$q = (T - T_s) k / x$$

Similarly, this quantity of heat must be transferred to the cooling medium from the food surface, so:

$$q = Ah_s(T_s - T_a)$$

where h_s is the surface heat transfer coefficient.

Eliminating T_s between the two equations gives:

$$q = (T - T_a)A \times 1/(1/h_s + x/k)$$

Since the same heat flow passing through the surface also passes through the frozen layer and is removed from the water as it freezes in the centre of the block:

$$A(T - T_a)/(1/h_s + x/k) = A\lambda\rho\ dx/dt$$

Therefore $(T - T_a)/(1/h_s + x/k) = \lambda\rho\ dx/dt$

$$dt(T - T_a) = \lambda\rho\ (1/h_s + x/k)dx$$

Now, if the thickness of the slab is a, the time taken for the centre of the slab at x = a/2 to freeze can be obtained by integrating from x = 0, to x = a/2 during which time t goes from t to t_f.

Therefore $t_f(T - T_a) = \lambda\rho(a/2h_s + a^2/8k)$

and $t_f = \dfrac{\lambda\rho}{(T - T_a)}(Pa/h_s + Ra^2/k)$

In his papers Plank derived his equation in more general terms and found that for brick-shaped solids the change is in numerical factors only.

A general equation can thus be written,

$$t_f = \frac{\lambda\rho}{(T - T_a)}(Pa/h_s + Ra^{\underline{o}}/k)$$

where P = 1/2, R = 1/8 for a slab; P = 1/4, R = 1/16 for an infinitely long cylinder, and P = 1/6, R = 1/24 for a cube or a sphere. Brick-shaped solids have values of P and R lying between those for slabs and those for cubes. Appropriate values of P and R for a brick-shaped solid can be obtained from the graph in Figure. In this figure, β_1 and β_2 are the ratios of the two longest sides to the shortest. It does not matter in what order they are taken.

Because the assumptions made in the derivation of Plank's equation lead to errors, which tend towards under-estimation of freezing times, more accurate predictions can be made if some allowances are made for this. One step is to use the total enthalpy changes from the initial to the final state of the product being frozen, that is to include the sensible heat changes both above and below the freezing temperature in addition to the latent heat. Even with this addition the prediction will still be about 20% or so lower than equivalent experimental measurements for brick shapes indicate. Adaptations of Plank's equation have been proposed which correspond better with experimental results, such as those in Cleland and Earle, but they are more complicated.

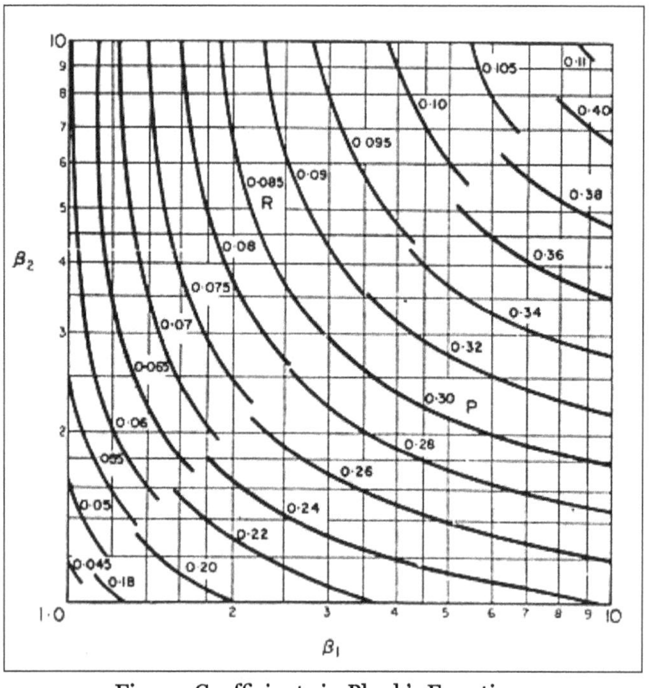

Figure: Coefficients in Plank's Equation.

Example: Freezing of a slab of meat.

If a slab of meat is to be frozen between refrigerated plates with the plate temperature at – 34°C, how long will it take to freeze if the slab is 10cm thick and the meat is wrapped in cardboard 1mm thick on either side of the slab? What would be the freezing time if the cardboard were not present? Assume that for the plate freezer, the surface heat transfer coefficient is 600Jm^{-2} s^{-1}°C^{-1}, the thermal conductivity of cardboard is 0.06Jm^{-1} s^{-1}°C^{-1} the thermal conductivity of frozen meat is 1.6J m^{-1} s^{-1}°C^{-1}, its latent heat is 2.56 × 105 Jkg^{-1} and density 1090 kg m^{-3}. Assume also that meat freezes at –2°C.

Conductance of cardboard packing = x/k = 0.001/0.06 = 0.017.

$1/h_s = x/k + 1/h_c = 0.017 + 1/600 = 0.019.$

In a plate freezer, the thickness of the slab is the only dimension that is significant. The case can be treated as equivalent to an infinite slab, and therefore the constants in Plank's equation are 1/2 and 1/8.

$$t_f = \frac{\lambda\rho}{(T - T_a)}\left(Pa/h_s + Ra^2/k\right)$$

$$= \frac{\left(2.56\times10^5\times1090\right)}{\left(-2-\{-34\}\right)}\times\left[\left(0.5\times0.1\times0.019\right)+\left(0.125\times\{0.1^2/1.6\}\right)\right]$$

$$= 1.51\times10^4\,s$$

$$= \underline{4.2h}$$

and with no packing $h_s = 600$

so that $1/h_s = 1.7 \times 10^{-3}$ and,

$$t_f = (2.56\times10^5 \times1090)/[-2 - \{-34\}]\times\left[\left(0.5 \times 0.1 \times 1.7 \times 10^{-3}\right)+\left(0.125 \times \{0.1\}^2/1.6\right)\right]$$

$$= 7.54\times10^3 s$$

$$= \underline{2.1h}$$

This estimate can be improved by adding to the latent heat of freezing, the enthalpy change above the freezing temperature and below the freezing temperature.

Above freezing, assume the meat with a specific heat of 3.22 × 103 J kg^{-1}°C starts at + 10°C and goes to –2 °C, needing 3.9 × 104 Jkg^{-1}.

Below the freezing temperature assume the meat goes from –2 °C to the mean of –2 °C and –34°C, that is –18°C, with a specific heat of 1.67×10^3 Jkg^{-1}°C needing 2.8×10^4 J kg^{-1}. This, with the latent heat of 2.56×10^5 gives a total of 3.23×10^5 J kg^{-1} and amended freezing times of: 4.2 × (3.23/2.56) = <u>5.3 h</u> and 2.1 × (3.23/2.56) = <u>2.6 h.</u>

If instead of a slab, cylinder, or cube, the food were closer to a brick shape then an additional 20% should be added.

Thus, with a knowledge of the thermal constants of a foodstuff, required freezing times can be estimated by the use of Plank's equation. The analysis using Plank's equation separates the total freezing time effectively into two terms, one intrinsic to the food material to be frozen, and the other containing the surface heat transfer coefficient which can be influenced by the process equipment. Therefore for a sensitivity analysis, it can be helpful to write Plank's equation in dimensionless form, substituting ΔH for λρ:

$$\frac{t\ h\ T}{a\ HP} = (1 + R\ Bi/P)$$

where $Bi = \underline{h}_a/k = h_s r/k$

This leads to defining an efficiency term:

$$\eta = \frac{RBi}{P} / \left(1 + \frac{RBi}{P}\right) = \frac{Bi}{P/R + Bi} = \frac{Bi}{4 + Bi}$$

(remembering that P/R = 4 for a slab),

In which η (eta) can be regarded as an efficiency of coupling of the freezing medium to the food varying from 1 for Bi → ∞ to 0 for Bi → 0. Taking an intrinsic freezing time t_f' for the case of unit driving force, ΔT = 1; then for general ΔT:

$$t_f = t_f'/(\eta\ \Delta T)$$

and can be used very readily to examine the influence of freezing medium temperature, and of surface heat transfer coefficient through the Biot number, on the actual freezing time. Thus the sensitivity of freezing time to process variations can be taken quickly into account.

Example: Freezing time of a carton of meat: Controllable factors.

Determine the intrinsic freezing time for the carton of Example. By putting the equation for the freezing time in the form evaluate the effect of,

- Changes in the temperature of the plates to −20°C, −25°C, and −30°C .
- Effect of doubling the thickness of the cardboard.
- Effect of decreased surface coefficients due to poor contact which drops the surface heat transfer coefficient to 100 $Jm^{-2}\ s^{-1}{}^{\circ}C^{-1}$.

$$Bi = \frac{h_s a}{k} = \frac{52.6 \times 0.1}{1.6} = 3.3$$

P/R = 4, and so η = 3.3/7.3 = 0.45

and from Example, calculated freezing time is 4.2h with driving force 32°C

Therefore t_f' = 4.2 x 0.45 x 32 = 60.5h

for (a)

$$t_f = 60.5/(0.45 \times 18) = \underline{7.5\text{ h for} - 20^{\circ}C}$$
$$60.5/(0.45 \times 23) = \underline{5.8\text{h for} - 25^{\circ}C}$$
$$60.5/(0.45 \times 28) = \underline{4.8\text{h for} - 30^{\circ}C}$$

for (b) x/k becomes 0.034, therefore $1/h_s$ = 0.034 + 0.0017

h_s = 28 $Jm^{-2}\ s^{-1}{}^{\circ}C^{-1}$

Bi becomes (28 × 0.1)/1.6 = 1.75

Therefore η becomes 0.30

so t_f = 60.5 × 1/0.30 + 1/32 = $\underline{6.3\text{ h}}$

for (c) 1/ h_s becomes 0.017 + 0.01 = 0.027,

h_s = 37 $Jm^{-2}\ s^{-1}{}^{\circ}C^{-1}$

Bi becomes (37 × 0.1)/1.6 = 2.3

Therefore η becomes 0.37

So t_f = 60.5 × 1/32 × 1/0.37 = 5.1 h

Cold Storage

For cold storage, the requirement for refrigeration comes from the need to remove the heat:

- Coming into the store from the external surroundings through insulation.

- From sources within the store such as motors, lights and people (each worker contributing something of the order of 0.5 kw).
- From the foodstuffs.

Heat penetrating thc walls can be estimated, knowing the overall heat transfer coefficients including the surface terms and the conductance of the insulation, which may include several different materials. The other heat sources require be considering and summing. Detailed calculations can be quite complicated but for many purposes simple methods give a reasonable estimate.

Drying

Drying is one of the oldest methods of preserving food. Primitive societies practiced the drying of meat and fish in the sun long before recorded history. Today the drying of foods is still important as a method of preservation. Dried foods can be stored for long periods without deterioration occurring. The principal reasons for this are that the microorganisms which cause food spoilage and decay are unable to grow and multiply in the absence of sufficient water and many of the enzymes which promote undesired changes in the chemical composition of the food cannot function without water.

Preservation is the principal reason for drying, but drying can also occur in conjunction with other processing. For example in the baking of bread, application of heat expands gases, changes the structure of the protein and starch and dries the loaf. Losses of moisture may also occur when they are not desired, for example during curing of cheese and in the fresh or frozen storage of meat, and in innumerable other moist food products during holding in air.

Drying of foods implies the removal of water from the foodstuff. In most cases, drying is accomplished by vaporizing the water that is contained in the food, and to do this the latent heat of vaporization must be supplied. There are, thus, two important process controlling factors that enter into the unit operation of drying:

- Transfer of heat to provide the necessary latent heat of vaporization.
- Movement of water or water vapour through the food material and then away from it to effect separation of water from foodstuff.

Drying processes fall into three categories:

- Air and contact drying under atmospheric pressure: In air and contact drying, heat is transferred through the foodstuff either from heated air or from heated surfaces and the water vapour is removed with the air.
- Vacuum drying: In vacuum drying, advantage is taken of the fact that evaporation of water occurs more readily at lower pressures than at higher ones. Heat transfer in vacuum drying is generally by conduction, sometimes by radiation.
- Freeze Drying: In freeze drying, the water vapour is sublimed off frozen food. The food structure is better maintained under these conditions. Suitable temperatures and pressures must be established in the dryer to ensure that sublimation occurs.

Theory of Drying Phase Diagram

Pure water can exist in three states, solid, liquid and vapour. The state in which it is at any time depends on the temperature and pressure conditions and it is possible to illustrate this on a phase diagram, as in figure.

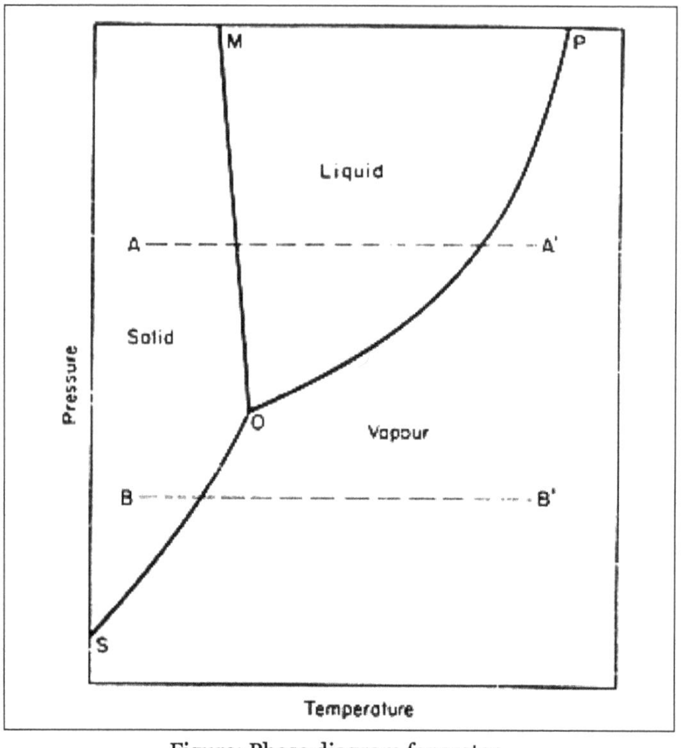

Figure: Phase diagram for water.

If we choose any condition of temperature and pressure and find the corresponding point on the diagram, this point will lie, in general, in one of the three labelled regions, solid, liquid, or gas. This will give the state of the water under the chosen conditions.

Under certain conditions, two states may exist side by side, and such conditions are found only along the lines of the diagram. Under one condition, all three states may exist together; this condition arises at what is called the triple point, indicated by point O on the diagram. For water it occurs at 0.0098°C and 0.64 kPa (4.8 mm of mercury) pressure.

If heat is applied to water in any state at constant pressure, the temperature rises and the condition moves horizontally across the diagram, and as it crosses the boundaries a change of state will occur. For example, starting from condition A on the diagram adding heat warms the ice, then melts it, then warms the water and finally evaporates the water to condition A'. Starting from condition B, situated below the triple point, when heat is added, the ice warms and then sublimes without passing through any liquid state.

Liquid and vapour coexist in equilibrium only under the conditions along the line OP. This line is called the vapour pressure/temperature line. The vapour pressure is the measure of the tendency of molecules to escape as a gas from the liquid. The vapour pressure/temperature curve for water is shown in figure, which is just an enlargement for water of the curve OP of figure.

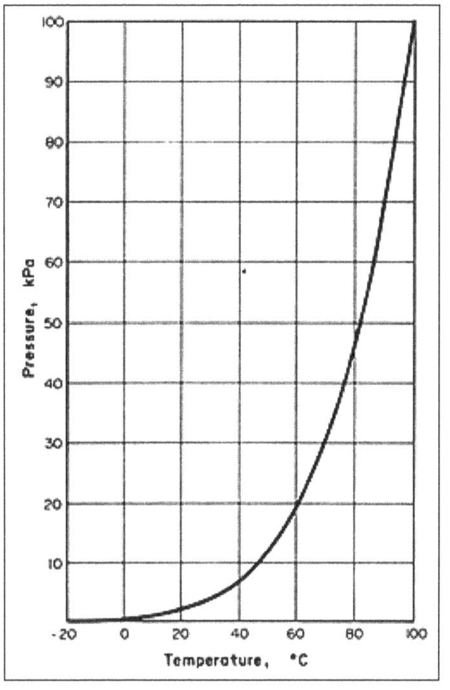

Figure: Vapour pressure/temperature curve for water.

Boiling occurs when the vapour pressure of the water is equal to the total pressure on the water surface. The boiling point at atmospheric pressure is of course 100°C. At pressures above or below atmospheric, water boils at the corresponding temperatures above or below 100°C, as shown in figure for temperatures below 100°C.

Latent Heat of Sublimation

Sublimation is a physical process in which a solid directly converts into a gaseous (vapor) state without going through a liquid state. The latent heat of sublimation at a particular temperature is the amount of heat required to convert a unit mass of solid into gas. For example, when ice sublimates into vapor, the amount of heat required at 0°C is estimated equal to 2,838 kJ/kg, which is the latent heat of sublimation of ice at 0°C. In the crystal growth of ice and snow in atmosphere, this process plays a dominant role. This process is opposite to the process of deposition.

Latent Heat of Vaporization

The enthalpy of vaporization, (symbol ΔH_{vap}) also known as the (latent) heat of vaporization or heat of evaporation, is the enthalpy change required to transform a given quantity of a substance from a liquid into a gas at a given pressure(often atmospheric pressure as in STP).

It is often measured at the normal boiling point of a substance; although tabulated values are usually corrected to 298 K, the correction is often smaller than the uncertainty in the measured value.

The heat of vaporization is temperature-dependent, though a constant heat of vaporization can be assumed for small temperature ranges and for reduced temperature $T_r << 1.0$. The heat of vaporization diminishes with increasing temperature and it vanishes completely at the critical temperature ($T_r = 1$) because above the critical temperature the liquid and vapour phases no longer exist, since the substance is a supercritical fluid.

Heat Requirements for Vaporization

The energy, which must be supplied to vaporize the water at any temperature, depends upon this temperature. The quantity of energy required per kg of water is called the latent heat of vaporization, if it is from a liquid, or latent heat of sublimation if it is from a solid. The heat energy required to vaporize water under any given set of conditions can be calculated from the latent heats given in the steam table, as steam and water vapour are the same thing.

Example: A food containing 80% water is to be dried at 100°C down to moisture content of 10%. If the initial temperature of the food is 21°C, calculate the quantity of heat energy required per unit weight of the original material, for drying under atmospheric pressure. The latent heat of vaporization of water at 100°C and at standard atmospheric pressure is 2257 kJ kg^{-1}. The specific heat capacity of the food is 3.8 kJ kg^{-1} $°C^{-1}$ and of water is 4.186 kJ kg^{-1} $°C^{-1}$. Find also the energy requirement/kg water removed.

Solution:

Calculating for 1 kg food:

- Initial moisture = 80% 800 g moisture are associated with 200 g dry matter.
- Final moisture = 10 %, 100 g moisture are associated with 900 g dry matter.

Therefore (100 x 200)/900 g = 22.2 g moisture are associated with 200 g dry matter. 1kg of original matter must lose (800 - 22) g moisture = 778 g = 0.778 kg moisture.

Heat energy required for 1kg original material = heat energy to raise temperature to 100°C + latent heat to remove water,

$= (100 - 21) \times 3.8 + 0.778 \times 2257$

$= 300.2 + 1755.9$

$= \underline{2056 \text{ kJ.}}$

Energy/kg water removed, as 2056 kJ are required to remove 0.778 kg of water,

$= 2056/0.778$

$= \underline{2643 \text{ kJ}}$

Steam is often used to supply heat to air or to surfaces used for drying. In condensing, steam gives up its latent heat of vaporization; in drying, the substance being dried must take up latent heat of vaporization to convert its liquid into vapour, so it might be reasoned that 1 kg of steam condensing will produce 1 kg vapour. This is not exactly true, as the steam and the food will in general be under different pressures with the food at the lower pressure. Latent heats of vaporization are slightly higher at lower pressures, as shown in Table. In practice, there are also heat losses and sensible heat changes which may require to be considered.

Table: Latent Heat and Saturation Temperature of Water.

Absolute pressure (kPa)	Latent heat of vaporization (kJ kg^{-1})	Saturation temperature (°C)
1	2485	7
2	2460	18
5	2424	33
10	2393	46
20	2358	60
50	2305	81
100	2258	99.6
101.35 (1 atm)	2257	100
110	2251	102
120	2244	105
200	2202	120
500	2109	152

Example: (Heat Energy in Vacuum drying).

Using the same material as in Example, if vacuum drying is to be carried out at 60°C under the corresponding saturation pressure of 20 kPa abs. (or a vacuum of 81.4 kPa), calculate the heat energy required to remove the moisture per unit weight of raw material.

Solution:

Heat energy required per kg raw material = heat energy to raise temperature to 60°C + latent heat of vaporization at 20 kPa abs.

$= (60 - 21) \times 3.8 + 0.778 \times 2358$

$= 148.2 + 1834.5$

$= \underline{1983 \text{ kJ.}}$

In freeze drying the latent heat of sublimation must be supplied. Pressure has little effect on the latent heat of sublimation, which can be taken as 2838 kJ kg^{-1}.

Example: (Heat Energy in Freeze drying).

If the foodstuff in the two previous Examples were to be freeze dried at 0°C, how much energy would be required per kg of raw material, starting from frozen food at 0°C?

Solution:

Heat energy required per kilogram of raw material= latent heat of sublimation,

$= 0.778 \times 2838$

$= \underline{2208 \text{ kJ.}}$

Heat Transfer in Drying

We have been discussing the heat energy requirements for the drying process. The rates of drying are generally determined by the rates at which heat energy can be transferred to the water or to the ice in order to provide the latent heats, though under some circumstances the rate of mass transfer (removal of the water) can be limiting. All three of the mechanisms by which heat is transferred - conduction, radiation and convection - may enter into drying. The relative importance of the mechanisms varies from one drying process to another and very often one mode of heat transfer predominates to such an extent that it governs the overall process.

As an Example, in air drying the rate of heat transfer is given by:

$$q = h_s A(T_a - T_s)$$

where q is the heat transfer rate in J s^{-1}, hs is the surface heat-transfer coefficient J m^{-2} s^{-1} $°C^{-1}$, A is the area through which heat flow is taking place, m^2, Ta is the air temperature and T_s is the temperature of the surface which is drying, °C.

To take another Example, in a roller dryer where moist material is spread over the surface of a heated drum, heat transfer occurs by conduction from the drum to the foodstuff, so that the equation is,

$$q = UA(T_i - T_s)$$

Where U is the overall heat-transfer coefficient, Ti is the drum temperature (usually very close to that of the steam), T_s is the surface temperature of the food (boiling point of water or slightly above) and A is the area of drying surface on the drum.

The value of U can be estimated from the conductivity of the drum material and of the layer of foodstuff. Values of U have been quoted as high as 1800 J m^{-2} s^{-1} $°C^{-1}$ under very good conditions and down to about 60 J m^{-2} s^{-1} $°C^{-1}$ under poor conditions.

In cases where substantial quantities of heat are transferred by radiation, it should be remembered that the surface temperature of the food may be higher than the air temperature. Estimates of surface temperature can be made using the relationships developed for radiant heat transfer although the actual effect of combined radiation and evaporative cooling is complex. Convection coefficients also can be estimated using the standard equations.

For freeze drying, energy must be transferred to the surface at which sublimation occurs. However, it must be supplied at such a rate as not to increase the temperature at the drying surface above the freezing point. In many applications of freeze drying, the heat transfer occurs mainly by conduction.

As drying proceeds, the character of the heat transfer situation changes. Dry material begins to occupy the surface layers and conduction must take place through these dry surface layers which are poor heat conductors so that heat is transferred to the drying region progressively more slowly.

Mass Transfer in Drying

In heat transfer, heat energy is transferred under the driving force provided by a temperature difference, and the rate of heat transfer is proportional to the potential (temperature) difference and to the properties of the transfer system characterized by the heat-transfer coefficient. In the same way, mass is transferred under the driving force provided by a partial pressure or concentration difference. The rate of mass transfer is proportional to the potential (pressure or concentration) difference and to the properties of the transfer system characterized by a mass-transfer coefficient.

Writing these symbolically, analogous to $q = UA\Delta T$, we have,

$$dw / dt = k'g\ A\Delta Y$$

Where w is the mass being transferred kg s^{-1}, A is the area through which the transfer is taking place, k'_g is the mass-transfer coefficient in this case in units of kg m^{-2} s^{-1}, and Y is the humidity difference in kg kg^{-1}.

Limitations: Unfortunately the application of mass-transfer equation is not as straightforward as heat transfer, one reason being because the movement pattern of moisture changes as drying proceeds. Initially, the mass (moisture) is transferred from the surface of the material and later, to an increasing extent, from deeper within the food to the surface and thence to the air. So the first stage is to determine the relationships between the moist surface and the ambient air and then to consider the diffusion through the food. In studying the surface/air relationships, it is necessary to consider mass and heat transfer simultaneously. Air for drying is usually heated and it is also a major heat-transfer medium. Therefore it is necessary to look carefully into the relationships between air and the moisture it contains.

Equilibrium Moisture Content

It is important to understand that there are limits to the amount of water which can be removed from a food material under any given conditions. The moisture content of wet material in equilibrium with air of given humidity and temperature is termed the equilibrium moisture content (EMC). The equilibrium moisture content of solid varies with the direction from which equilibrium is approached. When product losing moisture and attains EMC with surrounding, then the desorption isotherm is obtained. But if product gains moisture from the surrounding, and attains EMC, the adsorption isotherm is obtained. The characteristic hysteresis is produced by differences in adsorption and desorption of curve.

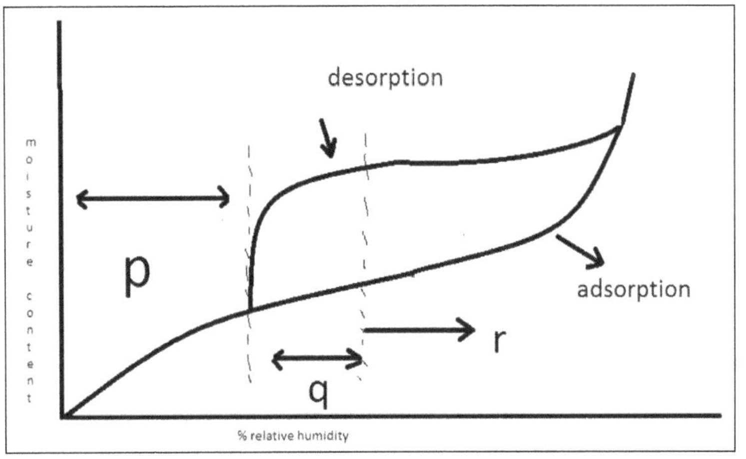

Figure: Adsorption-desorption isotherm.

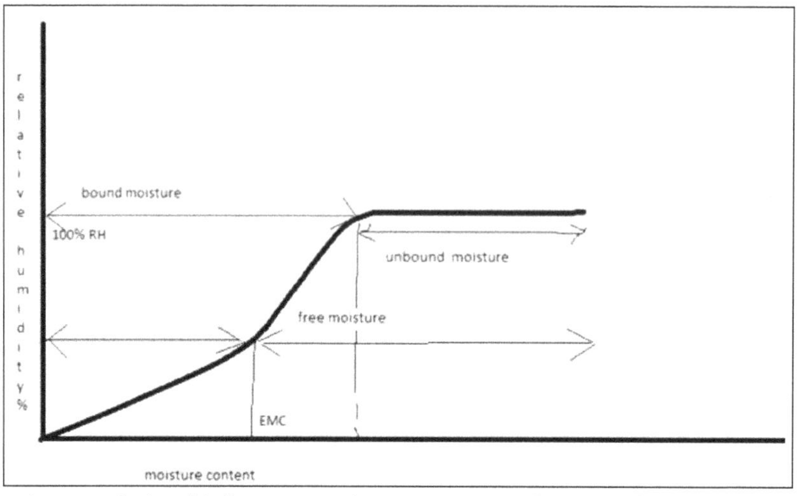

Figure: Relationship between moisture content and surrounding humidity.

A plot of EMC at given temperature versus the relative humidity is termed sorption isotherm. Figure shows two typical sorption isotherms. They are characterized by three distinct zones, which are discussed as below:

- Zone P: In zone P, the water is bound very strongly to sites and is unavailable for reaction. This moisture is referred to as monolayer moisture and it is strongly bound by adsorption forces. As shown in figure, it is observed that there is no distinction exists between the adsorption and desorption isotherm.
- Zone Q: In zone Q represents a transition zone. The water is bound in multiple layers; it is less strongly bound.
- Zone R: Water in zone R is even more loosely held. Water acts as a solvent and it can serve to mobilize the reactants.

Figure which represents plot % RH against the moisture content of food is different way of presenting the same information. The figure indicates three distinct moisture of product, which are defined as below:

- Bound Moisture: Bound moisture in a solid is determined from intersection of the equilibrium moisture content of the material with the 100% humidity line. The bound water exerts a vapour pressure lower than that of pure liquid at the same temperature. Substances containing bound water are called hygroscopic materials.
- Unbound Moisture: If a material contains more water than indicated by interaction with 100% humidity line, then this excessive moisture content is called unbound moisture and is held primarily in voids of the solid. It exerts a vapour pressure as high as that of ordinary water at the same temperature.
- Free Moisture: It is the moisture content of a material that can be removed by drying under the given temperature and humidity. It may include both bound and unbound moisture. Free moisture content obtained from the total average moisture content minus the equilibrium moisture content for the prevailing conditions of drying.

Psychrometry

The capacity of air for moisture removal depends on its humidity and its temperature. The study of relationships between air and its associated water is called psychrometry.

Humidity (Y) is the measure of the water content of the air. The absolute humidity is the mass of water vapour per unit mass of dry air and the units are therefore kg kg^{-1}. Absolute humidity is often called just 'humidity', as in this text. It is named absolute humidity or humidity ratio in charts.

Air is said to be saturated with water vapour at a given temperature and pressure if its humidity is a maximum under these conditions. If further water is added to saturated air, it must appear as liquid water in the form of a mist or droplets. Under conditions of saturation, the partial pressure of the water vapour in the air is equal to the saturation vapour pressure of water at that temperature.

The total pressure of a gaseous mixture, such as air and water vapour, is made up from the sum of the pressures of its constituents, which are called the partial pressures. Each partial pressure arises from the molecular concentration of the constituent and the pressure exerted by each gas is that which corresponds to the number of moles present and the total volume of the system. The partial pressures are added to obtain the total pressure.

Example: Partial pressure of water vapour.

If the total pressure of moist air is 100kPa (approximately atmospheric) and the humidity is measured as 0.03kg kg^{-1}, calculate the partial pressure of the water vapour.

The mole fraction of the water is the number of moles of water to the total number of moles (water + dry air).

The molecular weight of air is 29, and of water 18,

So the mole fraction of water = 0.03 /18 (1.00/29 + 0.03/18),

= 0.0017/(0.034 + 0.0017)

= 0.048.

Therefore the water vapour pressure = 0.048 x 100kPa = 4.8kPa.

The relative humidity (RH) is defined as the ratio of the partial pressure of the water vapour in the air (p) to the partial pressure of saturated water vapour at the same temperature (p_s).

Therefore:

$$RH = p / p_s$$

and is often expressed as a percentage = 100 p/p_s.

Example: Relative humidity.

If the air in Example is at 60°C, calculate the relative humidity.

From steam tables, the saturation pressure of water vapour at 600C is 19.9 kPa.

Therefore the relative humidity = p/p_s:

= 4.8/19.9

= 0.24 or 24%.

If such air were cooled, then when the percentage relative humidity reached 100% the air would be saturated and this would occur at that temperature at which $p = p_s = 4.8$ kPa.

Interpolating from the steam tables, or reading from the water vapour pressure/temperature graph, this occurs at a temperature of 32°C and this temperature is called the dew-point of the air at this particular moisture content. If cooled below the dew-point, the air can no longer retain this quantity of water as vapour and so water must condense out as droplets or a fog, and the water remaining as vapour in the air will be that corresponding to saturation at the temperature reached.

The humidity Y can therefore be related to the partial pressure p_w of the water in air vapour by the equation:

$$Y = 18\ p_w / \left[29\left(P - p_w\right)\right]$$

where P is the total pressure. In circumstances where p_w is small compared with P, and this is approximately the case in air/water systems at room temperatures, $Y \approx 18\ p_w / 29P$.

Corresponding to the specific heat capacity, c_p, of gases, is the humid heat, c_s of moist air. It is used in the same way as a specific heat capacity, the enthalpy change being the mass of dry air multiplied by the temperature difference and by the humid heat. The units are $Jkg^{-1}°C$ and the numerical values can be read off a psychrometric chart. It differs from specific heat capacity at constant pressure in that it is based only on the mass of the dry air. The specific heat of the water it contains is effectively incorporated into the humid heat which therefore is numerically a little larger than the specific heat capacity to allow for this.

Wet-bulb Temperatures

A useful concept in psychrometry is the wet-bulb temperature, as compared with the ordinary temperature, which is called the dry-bulb temperature. The wet-bulb temperature is the temperature reached by a water surface, such as that registered by a thermometer bulb surrounded by a wet wick, when exposed to air passing over it. The wick and therefore the thermometer bulb decreases in temperature below the dry-bulb temperature, until the rate of heat transfer from the warmer air to the wick is just equal to the rate of heat transfer needed to provide for the evaporation of water from the wick into the air stream.

Equating these two rates of heat transfer gives,

$$h_c A\left(T_a - T_s\right) = \lambda k_g{}' A\left(Y_s - Y_a\right)$$

where a and s denote actual and saturation temperatures and humidities; h_c is the heat transfer coefficient and k_g' the mass transfer coefficient from the air to the wick surface; λ is the latent heat of evaporation of water.

Therefore, $h_c / k_g' = \lambda(Y_s - Y_a)/(T_a - T_s)$

As the relative humidity of the air decreases, so the difference between the wet-bulb and drybulb temperatures, called the wet-bulb depression, increases and a line connecting wet-bulb temperature and relative humidity can be plotted on a suitable chart. When the air is saturated, the wet-bulb temperature and the dry-bulb temperature are identical.

Therefore if ($T_a - T_s$) is plotted against ($Y_s - Y_a$) remembering that the point (T_s, Y_s) must correspond to a dew-point condition, we then have a wet-bulb straight line on a temperature/humidity chart sloping down from the point (T_s, Y_s) with a slope of:

$$-(\lambda\, kg'/h_c)$$

A further important concept is that of the adiabatic saturation condition. This is the situation reached by a stream of water, in contact with the humid air. Both ultimately reach a temperature at which the heat lost by the humid air on cooling is equal to the heat of evaporation of the water leaving the stream of water by evaporation.

Under this condition with no heat exchange to the surroundings, the total enthalpy change (kJkg^{-1} dry air).

$$\begin{aligned} \Delta H &= c_s(T_a - T_s) + \lambda(Y_s - Y_a) = 0 \\ c_s &= -\lambda(Y_s - Y_a)/(T_a - T_s) \\ &= -h_c / k_g' \end{aligned}$$

where c_s is the humid heat of the air.

Now it just so happens, for the water/air system at normal working temperatures and pressures that for practical purposes the numerical magnitude of the ratio:

$$\frac{h_c}{c_s k_g'} \text{ (known as the Lewis number)} = 1$$

This has a useful practical consequence. The wet bulb line and the adiabatic saturation line coincide when the Lewis number = 1.

It is now time to examine the chart we have spoken about. It is called a psychrometric chart.

Psychrometric Charts

In the preceding discussion, we have been considering a chart of humidity against temperature, and such a chart is given in skeleton form on figure.

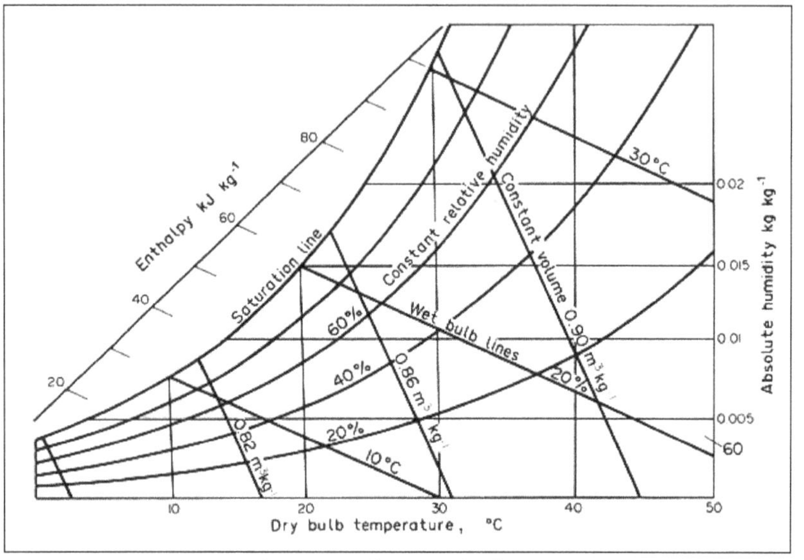

Figure: Psychrometric chart.

The two main axes are temperature (dry bulb) and humidity (absolute). The saturation curve (T_s, Y_s) is plotted on this dividing the whole area into an unsaturated and a two-phase region. Taking a point on the saturation curve (T_s, Y_s) a line can be drawn from this with a slope:

$$-\left(\lambda k_g'/h_c\right)=\left(-\lambda/c_s\right)$$

Running down into the unsaturated region of the chart (that "below" the saturation line). This is the wet bulb or adiabatic cooling line and a net of such lines is shown. Any constant temperature line running between the saturation curve and the zero humidity axis can be divided evenly into fractional humidities which will correspond to fractional relative humidities.

This discussion is somewhat over-simplified and close inspection of the chart shows that the axes are not exactly rectangular and that the lines of constant dry-bulb temperature are not exactly parallel. The reasons are beyond the scope of the present discussion but can be found in appropriate texts. The chart also contains other information whose use will emerge as familiarity grows. This chart can be used as the basis of many calculations. It can be used to calculate relative humidities and other properties.

Example: Relative humidity, enthalpy and specific volume of air.

If the wet-bulb temperature in a particular room is measured and found to be 20°C in air whose dry-bulb temperature is 25°C (that is the wet-bulb depression is 5°C) estimate the relative humidity, the enthalpy and the specific volume of the air in the room.

On the humidity chart follow down the wet-bulb line for a temperature of 20°C until it meets the dry-bulb temperature line for 25°C. Examining the location of this point of intersection with reference to the lines of constant relative humidity, it lies between 60% and 70%RH and about 4/10 of the way between them but nearer to the 60% line. Therefore the RH is estimated to be 64%. Similar examination of the enthalpy lines gives an estimated enthalpy of 57 kJ kg^{-1}, and from the volume lines specific volume of 0.862 m^3 kg^{-1}.

Once the properties of the air have been determined other calculations can easily be made.

Example: Relative humidity of heated air.

If the air in Example is then to be heated to a dry-bulb temperature of 40°C, calculate the rate of heat supply needed for a flow of 1000m^3 h^{-1} of this hot air for a dryer, and the relativc humidity of the heated air.

On heating, the air condition moves, at constant absolute humidity as no water vapour is added or subtracted, to the condition at the higher (dry bulb) temperature of 40°C. Reading from the chart at 40°C and humidity 0.0125kgkg^{-1}, the enthalpy is 73 kJ kg^{-1}, specific volume is 0.906m^3 kg^{-1} and RH 27%.

Mass of 1000m^3 is 1000/0.906 = 1104kg,

$\Delta H = (73 - 57) = 16kJkg^{-1}$.

So rate of heating required,

= 1104 x 16kJh^{-1}

= (1104 x 16)/3600kJs^{-1}

= 5Kw.

If the air is used for drying, with the heat for evaporation being supplied by the hot air passing over a wet solid surface, the system behaves like the adiabatic saturation system. It is adiabatic because no heat is obtained from any source external to the air and the wet solid, and the latent heat of evaporation must be obtained by cooling the hot air. Looked at from the viewpoint of the solid, this is a drying process; from the viewpoint of the air, it is humidification.

Example: Water removed in air drying.

Air at 60°C and 8%RH is blown through a continuous dryer from which it emerges at a temperature of 35°C. Estimate the quantity of water removed per kg of air passing, and the volume of drying air required to remove 20 kg water per hour.

Using the psychrometric chart, the inlet air condition shows the humidity ratio of the drying air to be 0.01kg kg^{-1} and its specific volume to be 0.96 m^3 kg^{-1}. Through the dryer, the condition of the air follows a constant wet-bulb line at around 27°C, so at 35°C its condition is a humidity ratio of 0.0207.

Water removed = (0.0207 - 0.010)

= 0.0107kgkg^{-1} of air.

So each kg, i.e. 0.96 m^3, of air passing will remove 0.0107kg water,

Volume of air to remove 20 kg h^{-1}

= (20/0.0107) × 0.96

= 1794m^{3}h^{-1}.

If air is cooled, then initially its condition moves along a line of constant humidity, horizontally on a psychrometric chart, until it reaches the saturation curve at its dew-point. Further cooling then proceeds down the saturation line to the final temperature, with water condensing to adjust the humidity as the saturation humidity cannot be exceeded.

Example: Relative humidity of air leaving a dryer.

The air emerging from a dryer, with an exit temperature of 45°C, passes over a surface which is gradually cooled. It is found that the first traces of moisture appear on this surface when it is at 40°C. Estimate the relative humidity of the air leaving the dryer.

On the psychrometric chart, the saturation temperature is 40°C and proceeding at constant humidity from this, the 45°C line is intersected at a point indicating: relative humidity = 76%.

In dryers, it is sometimes useful to reheat the air so as to reduce its relative humidity and thus to give it an additional capacity to evaporate more water from the material being dried. This process can easily be followed on a psychrometric chart.

Example: Reheating of air in a dryer.

A flow of 1800m^3 h^{-1} of air initially at a temperature of 18°C and 50% RH is to be used in an air dryer. It is heated to 140°C and passed over a set of trays in a shelf dryer, which it leaves at 60% RH. It is then reheated to 140°C and passed over another set of trays which it leaves at 60% RH again. Estimate the energy necessary to heat the air and the quantity of water removed per hour.

From the psychrometric chart, the humidity of the initial air is 0.0062kgkg^{-1}, specific volume is 0.834m3 kg^{-1}, and enthalpy 35kJkg^{-1}. Proceeding at constant humidity to a temperature of 140°C, the enthalpy is found (high temperature chart, Appendix 9(b)) to be 160 kJkg^{-1}. Proceeding along a wet-bulb line to an RH of 60% gives the corresponding temperature as 48°C and humidity as 0.045 kg kg^{-1}.

Reheating to 140°C keeps humidity constant and enthalpy goes to 268 kJ kg^{-1}.

Hence along a wet-bulb line to 60 % RH gives humidity of 0.082 kgkg^{-1}.

Total energy supplied in heating/reheating = ΔH

= 268 - 35

= 233kJkg^{-1}

Total water removed = ΔY

= 0.082 - 0.0062

= 0.0758kgkg^{-1}.

1800m^3 of air per hour = 1800/0.834

= 2158kgh^{-1}

= 0.6kgs^{-1}.

Energy taken in by air = 233 × 0.6kJs^{-1} = 140kW

Water removed in dryer = 0.6 × 0.0758

= 0.045kgs^{-1}

= 163kgh^{-1}

Exit temperature of air (from chart) = 60°C

Consideration of psychrometric charts, and what has been said about them, will show that they can be used for calculations focused on the air, for the purposes of air conditioning as well as for drying.

Example: Air conditioning.

In a tropical country, it is desired to provide processing air conditions of 15°C and 80% RH. The ambient air is at 31.5°C and 90% RH. If the chosen method is to cool the air to condense out enough water to reduce the water content of the air sufficiently, then to reheat if necessary, determine the temperature to which the air should be cooled, the quantity of water removed and the amount of reheating necessary. The processing room has a volume of 1650 m^3 and it is estimated to require six air changes per hour.

Using the psychrometric chart (normal temperatures):

- Initial humidity is 0.0266 kg kg^{-1}.
- Final humidity is 0.0085 kg kg^{-1}.
- Saturation temperature for this humidity is 13°C.
- Therefore the air should be cooled to 13°C.
- At the saturation temperature of 13°C, the enthalpy is 33.5 kJ kg^{-1}.
- At the final conditions, 15°C and 80 % RH, the enthalpy is 37 kJ kg^{-1}.
- and the specific volume of air is 0.827 m3 kg^{-1}.
- Assuming that the air changes are calculated at the conditions in the working space.
- Mass of air to be conditioned = (1650 × 6)/0.827

  = 11,970kgh^{-1}
- Water removed per kg of dry air ΔY:

  = 0.0266 - 0.0085

  = 0.018kgkg^{-1}.
- Mass of water removed per hour:

  = 11,970 x 0.018

  = 215kgh^{-1}

- Reheat required $\Delta H = (37 - 33.5)$:

  $= 3.5 kJkg^{-1}$

- Total reheat power required = $11{,}970 \times 3.5$:

  $= 41{,}895 kJh^{-1}$

  $= 11.6 kJs^{-1} = 11.6 kW$.

Measurement of Humidity

Methods depend largely upon the concepts that have been presented in the preceding sections, but because they are often needed it seems useful to set them out specifically. Instruments for the measurement of humidity are called hygrometers.

- Wet- and dry-bulb thermometers: The dry-bulb temperature is the normal air temperature and the only caution that is needed is that if the thermometer bulb, or element, is exposed to a surface at a substantially higher or lower temperature the possibility of radiation errors should be considered. A simple method to greatly reduce any such error is to interpose a radiation shield, e.g. a metal tube, which stands off from the thermometer bulb 1cm or so and prevents direct exposure to the radiation source or sink. For the wet bulb thermometer, covering the bulb with a piece of wicking, such as a hollow cotton shoelace of the correct size, and dipping the other end of the wick into water so as to moisten the wet bulb by capillary water flow, is adequate. The necessary aspiration of air past this bulb can be effected by a small fan or by swinging bulb, wick, and water bottle and all through the air, as in a sling psychrometer. The maximum difference between the two bulbs gives the wet-bulb depression and a psychrometric chart or appropriate tables will then give the relative humidity.
- Dew-point meters: These measure the saturation or dew-point temperature by cooling a sample of air until condensation occurs. The psychometric chart or a scale on the instrument is then used to give the humidity. For example, a sample of air at 20°C is found to produce the first signs of condensation on a mirror when the mirror is cooled to 14°C. The chart shows by moving horizontally across, from the saturation temperature of 14°C to the constant temperature line at 20°C, that the air must have a relative humidity of 69%.
- The hair hygrometer: Hairs expand and contract in length according to the relative humidity. Instruments are made which give accurately the length of the hair and so they can be calibrated in humidities.
- Electrical resistance hygrometers: Some materials vary in their surface electrical resistance according to the relative humidity of the surrounding air. Examples are aluminium oxide, phenol formaldehyde polymers, and styrene polymers. Calibration allows resistance measurements to be interpreted as humidity.
- Lithium chloride hygrometers: In these a solution of lithium chloride is brought to a temperature such that its partial pressure equals the partial pressure of water vapour in the air. The known vapour pressure/temperature relationships for lithium chloride can then be used to determine the humidity of the air.

Methods of Drying

According to the mode of heat transfer, drying methods can be divided into (a) Conduction drying; (b) convection drying; and (c) radiation drying. There are other methods of drying also, namely, dielectric drying, chemical or sorption drying, vacuum drying, freeze-drying, etc.

Convection Drying

In convection drying, the drying agent (hot gases) in contact with the wet solid is used to supply heat and carry away the vaporized moisture, and the heat is transferred to the wet solid mainly by convection. The characteristics of convection drying are as follows:

- Drying is dependent upon the heat transfer from the drying agent to the wet material, the former being the carrier of vaporized moisture.
- Steam heated air, direct flue gases of agricultural waste, etc., can be used as drying agents.
- Drying temperature varies widely.
- At gas temperatures below the boiling point, the vapour content of the gas affects the drying rate and the final moisture content of the solid.
- If the atmospheric humilities are high, natural air drying needs dehumidification.
- Fuel consumption per kilogram of moisture evaporated is always higher than that of conduction drying.

Convection drying is most popular in grain drying. It can be carried out either continuously or batch-wise. Continuous tray dryers, continuous sheeting dryers, pneumatic conveying dryers, rotary dryers, and tunnel dryers come under the continuous system, whereas tray and compartment dryers and batch through circulation dryers are batch dryers. Convection drying can be further classified as follows:

Pneumatic of fluidized bed drying: When the hot gas (drying agent) is supplied at a velocity higher than the terminal velocity of the wet solid, the drying of the wet solid occurs in a suspended or fluidized state. This phenomenon is known as fluidized bed drying.

Drying may be carried out in a semi-suspended state or spouted bed condition also. Generally, the convection drying is conducted under ordinary state, i.e., drying agent is supplied at a velocity much lower than the terminal velocity of the wet material.

Conduction Drying

When the heat for drying is transferred to the wet solids mainly by conduction through a solid surfaces the phenomenon is known as conduction or contact drying. In this method, conduction is the principal mode of heat transfer and the vaporized moisture is removed independently of the heating media. Conduction drying is characterized by:

- Heat transfer to the wet solid takes place by conduction through a solid surface, usually metallic. The source of heat may be hot water, steam, flue gases, hot oil etc.

- Surface temperatures may vary widely.
- Contact dryers can be operated under low pressure and inert atmosphere.
- Dust and dusty materials can be removed very effectively.

Conduction drying can be carried out either continuously or batch wise. Cylinder dryers, drum dryers, steam tube rotary dryers are some of the continuous conduction dryers. Vacuum tray dryers, Freeze dryers and agitated pan dryers are the Examples of batch conduction dryers.

In an industry as diversified and extensive as the food industry, it would be expected that a great number of different types of dryer would be in use. The principles of drying may be applied to any type of dryer, but it should help the understanding of these principles if a few common types of dryers are described. The major problem in calculations on real dryers is that conditions change as the drying air and the drying solids move along the dryer in a continuous dryer, or change with time in the batch dryer.

- Tray Dryers: In tray dryers, the food is spread out, generally quite thinly, on trays in which the drying takes place. Heating may be by an air current sweeping across the trays, by conduction from heated trays or heated shelves on which the trays lie, or by radiation from heated surfaces. Most tray dryers are heated by air, which also removes the moist vapours.
- Tunnel Dryers: These may be regarded as developments of the tray dryer, in which the trays on trolleys move through a tunnel where the heat is applied and the vapours removed. In most cases, air is used in tunnel drying and the material can move through the dryer either parallel or counter current to the air flow. Sometimes the dryers are compartmented, and cross-flow may also be used.
- Roller or Drum Dryers: In these the food is spread over the surface of a heated drum. The drum rotates, with the food being applied to the drum at one part of the cycle. The food remains on the drum surface for the greater part of the rotation, during which time the drying takes place, and is then scraped off. Drum drying may be regarded as conduction drying.
- Fluidized Bed Dryers: In a fluidized bed dryer, the food material is maintained suspended against gravity in an upward-flowing air stream. There may also be a horizontal air flow helping to convey the food through the dryer. Heat is transferred from the air to the food material, mostly by convection.
- Spray Dryers: In a spray dryer, liquid or fine solid material in slurry is sprayed in the form of a fine droplet dispersion into a current of heated air. Air and solids may move in parallel or counter flow. Drying occurs very rapidly, so that this process is very useful for materials that are damaged by exposure to heat for any appreciable length of time. The dryer body is large so that the particles can settle, as they dry, without touching the walls on which they might otherwise stick. Commercial dryers can be very large, 10m diameter, 20m high.
- Pneumatic Dryers: In a pneumatic dryer, the solid food particles are conveyed rapidly in an air stream, the velocity and turbulence of the stream maintaining the particles in suspension. Heated air accomplishes the drying and often some form of classifying device is included in the equipment. In the classifier, the dried material is separated, the dry material passes out as product and the moist remainder is recirculate for further drying.

- Rotary Dryers: The foodstuff is contained in a horizontal inclined cylinder through which it travels, being heated either by air flow through the cylinder, or by conduction of heat from the cylinder walls. In some cases, the cylinder rotates and in others the cylinder is stationary and a paddle or screw rotates within the cylinder conveying the material through.
- Trough Dryers: The materials to be dried are contained in a trough-shaped conveyor belt, made from mesh, and air is blown through the bed of material. The movement of the conveyor continually turns over the material, exposing fresh surfaces to the hot air.
- Bin Dryers: In bin dryers, the foodstuff is contained in a bin with a perforated bottom through which warm air is blown vertically upwards, passing through the material and so drying it.
- Belt Dryers: The food is spread as a thin layer on a horizontal mesh or solid belt and air passes through or over the material. In most cases the belt is moving, though in some designs the belt is stationary and the material is transported by scrapers.
- Vacuum Dryers: Batch vacuum dryers are substantially the same as tray dryers, except that they operate under a vacuum, and heat transfer is largely by conduction or by radiation. The trays are enclosed in a large cabinet, which is evacuated. The water vapour produced is generally condensed, so that the vacuum pumps have only to deal with non-condensable gases. Another type consists of an evacuated chamber containing a roller dryer.
- Freeze Dryers: The material is held on shelves or belts in a chamber that is under high vacuum. In most cases, the food is frozen before being loaded into the dryer. Heat is transferred to the food by conduction or radiation and the vapour is removed by vacuum pump and then condensed. In one process, given the name accelerated freeze drying, heat transfer is by conduction; sheets of expanded metal are inserted between the foodstuffs and heated plates to improve heat transfer to the uneven surface and moisture removal. The pieces of food are shaped so as to present the largest possible flat surface to the expanded metal and the plates to obtain good heat transfer. A refrigerated condenser may be used to condense the water vapour.

Various types of dryers are illustrated in figure.

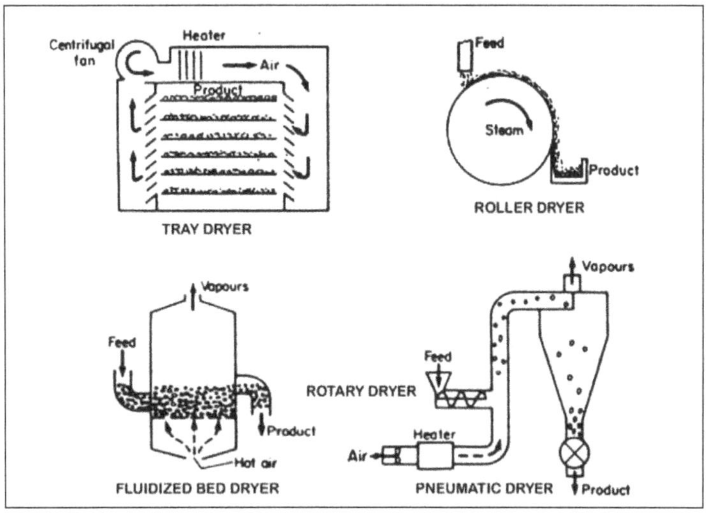

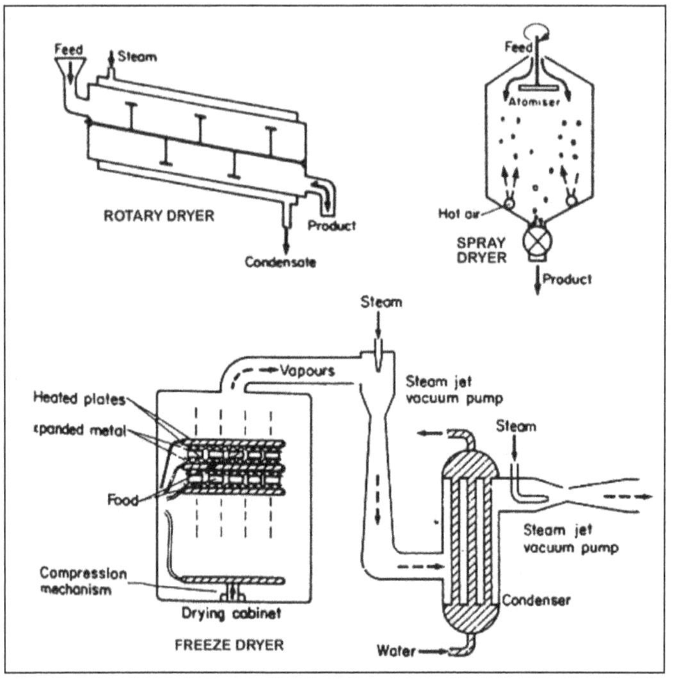

Figure: Dryers

Evaporation

Evaporation is a common unit operation in food processing industries in which partial removal of water takes place from an aqueous solution. The removal of water in the form of vapour is achieved by the process of vaporization or boiling of aqueous solution. Evaporation is confined to the preservation of liquid foods by concentrating them as a result, microbiological damage is checked and shelf life increased. One more advantage is the reduction in mass and volume of material, resulting in reduced cost of packaging, storage and transportation. The application of evaporation in food industries includes concentration of aqueous solution of sugar, milk, sodium chloride, fruit juices etc. In all the cases, the concentrated solutions are the required products and removed water is discarded. But, in processes like desalination, water is the required product. Sometimes, evaporation is used as an intermediate operation mainly before drying, particularly in spray drying. This helps increased economy of dryer, as the removal of huge amount of water in processes like spray drying is a costly affair.

While selecting the adaptability of evaporation in different food products, one cannot overlook the physico-chemical changes that take place due to processing factors like exposure of products to high temperature and pressure conditions, which reduces the product quality in terms of vitamin and essential volatile components. One major cause of concern is the concentration of liquid solution. As the evaporation proceeds, the relatively dilute solution concentrates; as a result heat transfer coefficients reduce and load on the pumping system increases. The boiling point of solution rises as the solute concentration increases, as a consequence, the temperature difference which is the driving force in heat transfer between the steam and boiling point of aqueous solution decreases. To keep boiling point of liquid food as low as possible, the evaporators are operated under vacuum, because reduced pressure reduces the boiling point temperature. As the boiling point temperature reduces, the thermal degradation to foods minimizes. One disadvantage of low temperature operation would be increased viscosity of solution, causing reduced heat transfer.

The solubility of solids in a solution decreases with increase in concentration. The insoluble solids called scales are deposited on the wall of the evaporator, which reduces the overall heat transfer coefficients. Fouling may occur which causes microbiological damage to the products. Solubility is related to the crystallization process, the soluble solids at high temperature (super saturation) when cooled crystallization occurs, which is undesirable in concentration process as it may cause blockage to the evaporator.

Evaporators and their Types

The equipments used to evaporate water from an aqueous solution are called evaporators. The schematic diagram of a single effect evaporator is given in figure. The system consists of the following components:

- An evaporation chamber.
- A heat exchanger to supply heat to vaporize water.
- A condenser to suck the vapour from the evaporation chamber.
- A steam jet ejector to remove non condensable gases.

The main components of the evaporator are the evaporation chamber and the heat exchanger which occupies a part of evaporation chamber. The rate of evaporation in an evaporator is determined by the amount of heat transferred to the solution by the heat exchanger. The steam gives off its latent heat to the surface of the exchanger which is in contact with the feed solution. The latent heat at atmospheric pressure is 2257 kJ/kg. Since, steam enters at superheated condition the heat supplied is more than 2257 kJ/kg steam. If there are no losses, same amount of heat is available to the solution to reach to the boiling point at prevailing pressure inside the evaporator. But, practically the heat losses cannot be avoided. There are many factors like steam saturation level, temperature and pressure at which steam is supplied, heat transfer area, the feed solution temperature, viscosity of feed, the fouling factor etc. decides the rate of heat transfer. Steam after it gives off all the heat is collected in the form of condensate liquid. The evaporation chamber or vapour chamber is the largest visible part of evaporator. This allows the separation of vapour from feed solution. The vapour temperature is the saturation steam temperature at the absolute pressure inside the chamber. If the boiling point rise is neglected the boiling temperature of feed, vapour temperature and product temperature are same i.e. they are in thermal equilibrium. The vapour formed must be removed from the vapour chamber to allow the space for evaporation, otherwise the accumulated vapour would increase the absolute pressure inside the chamber, as a result the boiling point will rise and the temperature difference will decrease.

Surface condenser is used to recover vapour from the chamber like the cases where some valuable components are driven out by vapour and are collected from subsequent operations. They are very expensive, so seldom used in food industries. The common condensers are direct contact water cooled condensers. The vapour enters a water spray chamber which is connected to a barometric leg. The barometric leg full of water balances the pressure inside the chamber and atmosphere. It maintains vacuum in the vapour chamber intact. The temperature of the water-condensate mixture must be below the vapour temperature to allow the vapour flow from the chamber to the condenser. In order to check the flow of feed solution with the vapour baffle plates are installed near the vapour outlet. Some amount of vapour may not condensate in water spraying chamber. They are removed by a steam jet ejector, where high pressure steam is passed through a narrow opening and the non-condensable vapour is sucked along with the high velocity steam due to low pressure generation. The evaporator is also called as 'effect'.

The orientation of heat exchanging area i.e. steam tubes on the chamber and the means of agitation/ circulation provided evolved different types of evaporators. All evaporators are continuous types.

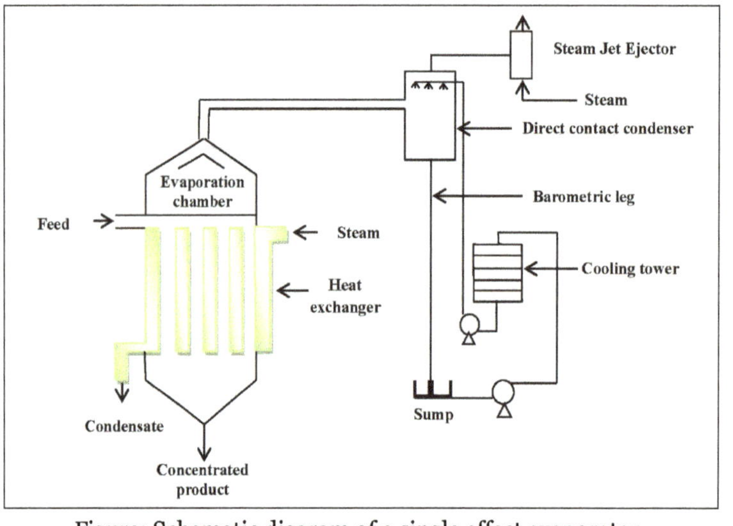

Figure: Schematic diagram of a single effect evaporator.

Horizontal Short Tube Evaporator

The steam tubes or bundles are oriented horizontally in the evaporator. Steam enters at one end of the tube, gives up latent heat and condensate leaves out from the other end. The feed liquid covers the steam bundles from outside surface of the tubes where heat transfer takes place. Since the circulation of feed is poor, low viscosity solutions can only be used in this case. Heat transfer coefficient of feed must be high to compensate non agitation of standing feed in the vapour chamber. The product should be non-corrosive to the pipes.

Vertical Short Tube Evaporator

The tubes are arranged vertically inside which feed solution passes. The vapour condensed outside the tube bundles. The lighter liquid because of boiling and decreased density moves up and heavier liquid moves down through a central larger pipe. Heat transfer increases due to natural circulation of liquid. Viscous liquids cannot climb on the tubes, so viscous products are not evaporated in vertical short tube evaporators. The length of tubes varies from 0.5 to 2.0 m with diameter of tube ranging between 25 to 75 mm. The central tube occupies 25 to 40% of the total tube area. This evaporator is generally use for evaporation of sugar and salt solutions.

Vertical Long Tube Evaporator

Vertical long tube evaporator is a little modification of vertical short tube evaporator. The liquid tubes are 3 to 10 m long. Central downstream feed tube is not required since the liquid is fed from the bottom and climbs up the tubes once. As the liquid passes through the narrow tubes, the bubbles formed due to evaporation of liquid drags the liquid and don't allow liquid movement downward. Agitation and circulation provided by vapour bubbles is high, so the heat transfer in this case is high. The contact time is very less due to high velocity liquid, thereby thermal damage to the product is very less. The liquid vapour mixture enters a separation chamber, where baffle plates are provided to facilitate vapour separation. The concentrated solution is collected from the bottom of the separator or if desired concentration is not achieved, is recycled with the fresh feed. Milk is concentrated in this evaporator mostly.

Falling Film Evaporator

The feeding position on the long tube evaporators can be provided at the top instead of bottom feeding. The long tube evaporator with top feeding is called as falling film evaporator. The vapour formed inside the tube descends as a jet at high velocity. Heat sensitive foods like fruit juices are concentrated in falling film evaporators. The descending liquid moves at higher velocity compared to ascending liquid, because ascending liquid moves against the direction of gravitational force. So, the residence time in falling film is very less (in the range of 5 to 10 s) compared to ascending long tube evaporator.

Forced Circulation Evaporator

The circulation velocity can be achieved by externally mounted feed pump. The tube length in forced circulation evaporator need not be long to achieve film velocity. The circulation velocity of liquid generally varies from 2 to 6 m/s inside the tube. High viscous liquids can be evaporated in this evaporator as adequate force is applied by the pump to move the viscous liquid in the tube.

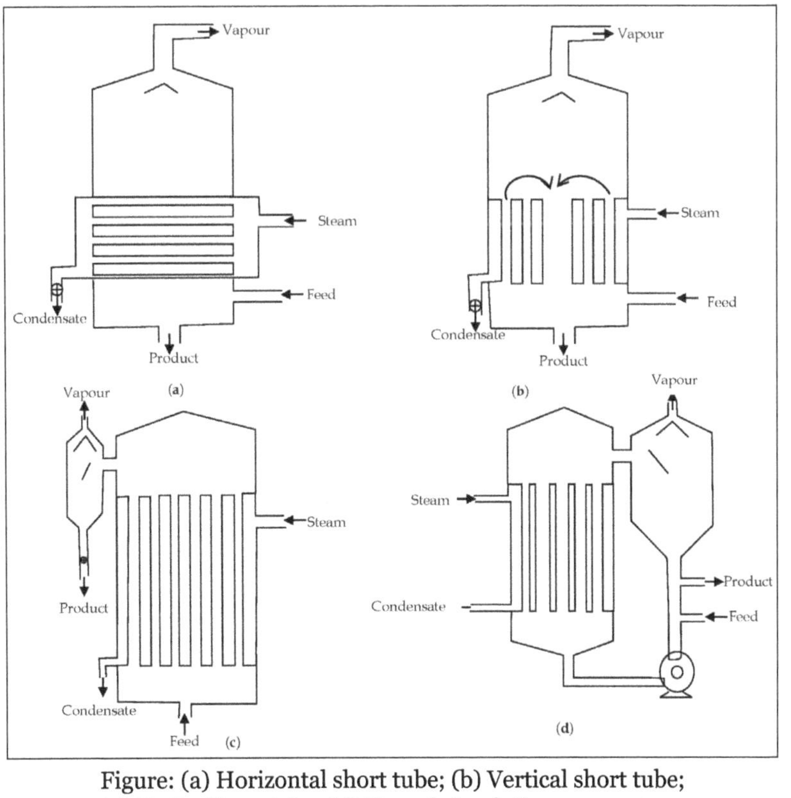

Figure: (a) Horizontal short tube; (b) Vertical short tube; (c) Vertical long tube; (d) Forced circulation evaporators.

Apart from these types of evaporators, there are other evaporators like scrapped surface, open pan, agitated film evaporators used in food industries. However, the falling film evaporator is most common among all the evaporators.

Heat Transfer Calculation in Single Effect Evaporator

The schematic diagram of a single effect evaporator is given in figure. The feed solution enters the evaporator at the rate of F kg/h, the solute concentration x_f, temperature of feed T_f and enthalpy H_f. The concentrated solution leaves out at the rate of C kg/h at T_c having solute concentration x_c and enthalpy H_c. The rate of vapour removal is V kg/h at vapour condition T_v and H_v. Since the vapour is in pure form the solute concentration xv is zero. Steam enters the condensation chamber in saturated form at S kg/h which has temperature and enthalpy of T_s and H_s respectively. Since vapour condenses and gives up only the latent heat, condensate leaves at the same rate and temperature, but the enthalpy is reduced to h_s. Since vapour and liquid are in equilibrium state, the temperature inside the evaporator, T_e, T_v and T_c are same. The boiling temperature T_e corresponds to saturation vapour pressure of P_e inside the evaporator.

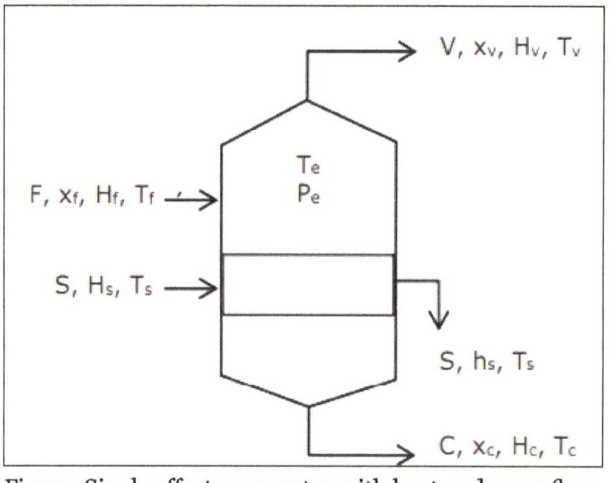

Figure: Single effect evaporator with heat and mass flow.

Since, the system is in steady state, mass entering is equal to mass leaving. Total mass balance:

$$F = C + V$$

Total solute/solid balance:

$$Fx_f = Cx_c + Vx_v$$

Since $x_v = 0, Fx_f = Cx_c$

Total enthalpy balance:

heat in steam + heat in feed = heat in product + heat in vapour + heat in condensate

$$FH_f + SH_s = CH_c + VH_v + Sh_s$$

But, Latent heat of condensation, $\lambda = H_s - h_s$

Substituting equation $\lambda = H_s - h_s$ into equation $FH_f + SH_s = CH_c + VH_v + Sh_s$, we get:

$$FH_f + S\lambda = CH_c + VH_v$$

Total heat transferred to the evaporator is $S\lambda$. This total heat transferred must be equal to the heat gain.

$$Q = S\lambda = UA(T_S - T_e)$$

Where, U is overall heat transfer coefficient and A is the surface area of heat transfer of evaporator. Once A is determined we can design the evaporator size. If the heat capacity of feed and product are known, we can find out the enthalpies.

Economy of Evaporator

Steam Economy (E_s)

The capacity of evaporator is defined as the amount of vapour evaporated i.e. V kg/h. The steam economy is determined by evaluating how much of steam S kg/h is consumed against V kg/h vapour evaporation.

So, steam economy $E_S = \frac{V}{S}$

Steam economy mostly depends on how best we can use the heat in evaporation system. But, in single effect evaporator it is seen that, almost 1 kg steam is required to form 1 kg vapour, i.e. steam economy is one. Maximum heat is lost through the vapour. There are some techniques by which we can improve the economy by utilizing the wasting heat of vapour. They are:

- Multiple effect evaporators,
- Mechanical vapour recompression,
- Thermal vapour recompression.

Multiple Effect Evaporators

It is already been mentioned, the single unit of evaporator is called an 'effect'. If the evaporator units of identical size are used in series, the evaporation system is called as multiple effect evaporator. The vapour of first effect is used as the heat source in the second effect and the concentrated liquid of first effect is fed to second effect for further concentration. This arrangement is maintained in subsequent effects also. The steam economy is increased as the steam is applied once and vapour formation in all effects is huge. So, overall economy of evaporator is increased. In triple effect evaporator, with the use of 1 kg steam, almost 3 kg vapour is formed, i.e. number of effect we use, same number of vapour evaporated. In food processing industries generally triple effect evaporators are used. Since, the vapour temperature of the first effect is the heating temperature for second effect, the temperature difference between the heating temperature of vapour and boiling temperature of the liquid decreases. To maintain the temperature difference high, the second effect and subsequent effects are operated at vacuum in gradual manner. If we see closely, the advantage of steam economy is overshadowed by the maintenance cost of the additional evaporators. The multiple effect evaporators are classified according to the direction of feed and steam.

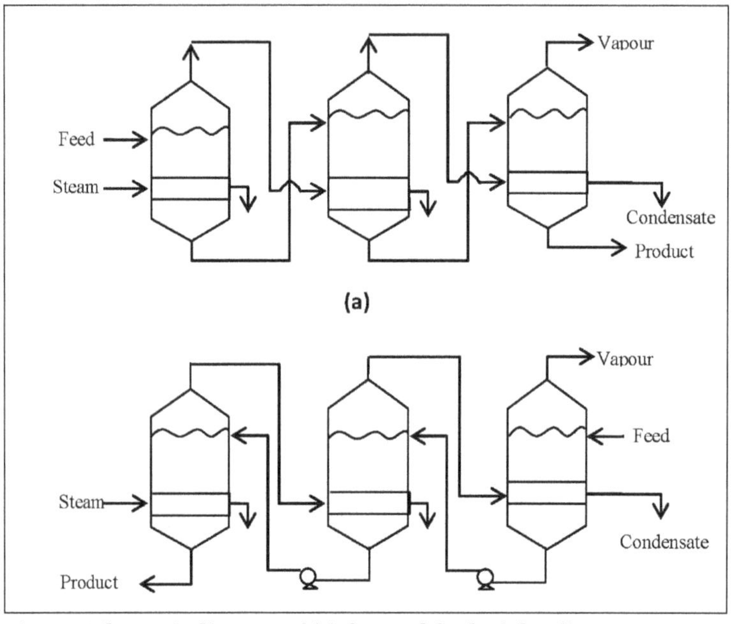

Figure: Schematic diagram of (a) forward feed triple effect evaporator; (b) backward feed triple effect evaporator.

a) Forward feed: Figure shows a schematic diagram of forward feed triple effect evaporator. The steam is fed to the first effect, the vapour formed is passed to the calandria of second effect and the vapour formed in the second effect is passed to third effect. The feed solution is fed to first effect, the concentrated solution of first effect is fed to the second effect and the further more concentrated solution of second effect is fed to the third effect. Ultimately, the required concentration product is collected from the third effect and vapour is vented out which is at very low temperature. No pumps are required to move the concentrated liquid from one effect to subsequent effect since product flow from high pressure to low pressure effects. However, a pump is required to feed the liquid to first effect since the first effect is at atmospheric pressure. These types of evaporators are used when the feed temperature is too high to carry over the enthalpy to subsequent effects.

b) Backward feed: In backward feed evaporator the feed is fed from the third effect and the steam is supplied from the first effect in the opposite direction. By doing so, the most concentrated solution is exposed to high temperature in first effect and the dilute solution is exposed to lower temperature in third effect. The more viscous solution when exposed to high heat the flow ability is maintained by decreasing its viscosity, thereby the overall heat transfer is not hampered. One disadvantage of this system is the use of pumps to pass the feed from low pressure to high pressure effects, thus increased cost of system. More viscous products use backward feed evaporators.

c) Mixed feed: In this mixed feed system, the feed is fed from the intermittent effect and passed in the forward direction parallel to the flow of steam. The relatively concentrated solution is then pumped to the left over first few evaporators in the backward flow fashion. The use of some pumps is avoided unlike backward feed evaporators, and at the same time more concentrated product is achieved.

Heat Transfer in Multiple Effect Evaporator

The driving force of heat transfer in the evaporator is the temperature difference. In multiple effect evaporator, there must exist some temperature difference in each effect i.e. the boiling temperature of first effect T_1 °C should be greater than the boiling temperature T_2 °C and T_3 °C of second and third effects respectively. The reduced boiling point temperature is achieved with reduction in pressure. Thus, $P_1 > P_2 > P_3$.

Now, if the boiling point rise is negligible, the total temperature drop will be,

$$\Delta T_{overall} = T_s - T_v$$

Where T_s and T_v are the temperatures of saturated steam and the saturated vapour of the last effect. The detailed material balance can be achieved with some assumptions made as follows:

1. The latent heat of evaporation does no change markedly with pressure drop.

2. Heat needed to heat up the feed up to boiling temperature is negligible compared to heat required for vapour formation.

3. Heat losses and boiling point elevation are negligible.

With these assumptions we can say, the amount of vapour produced in effect are same.

So, $S = V_1 = V_2 = V_3$

From the equation above we can conclude that the heat transfer in each effect is same.

$$Q_1 = Q_2 = Q_3$$

Or, $U_1A_1(T_s - T_1) = U_2A_2(T_1 - T_2) = U_3A_3(T_2 - T_3)$

The total heat transfer $Q = U_{overall} A \Sigma \Delta T$

Where, $U_{overall} = \dfrac{1}{1/U_1 + 1/U_2 + 1/U_3}$

Let us assume the areas and the overall heat transfer coefficients of all the effect are same.

$$A_1 = A_2 = A_3$$
$$U_1 = U_2 = U_3$$

So, the total vapour evaporated from the system is,

$$V = \frac{UA(T_s - T_3)}{\lambda}.$$

One conclusion we draw from the above relations is that, the single effect evaporator with its overall heat transfer coefficient same as that of multiple effect, the area same as the area of one effect and the temperature difference same as the overall temperature difference of multiple effect will have same capacity Q. Here it is evident that the evaporation capacity is independent of number of effect.

Mechanical Vapour Recompression (MVR) System

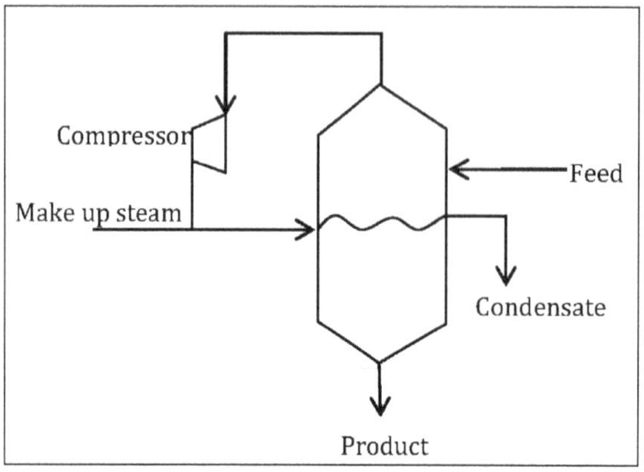

The enthalpy of vapour which is generally wasted is reused as steam in the evaporator. The reduced enthalpy of vapour from that of the live steam can be made up by compressing the vapour using a mechanical compressor. The principle of increased enthalpy with pressure is depicted in Mollier diagram. The inevitable loss of vapour is compensated with the supply of same amount of makeup live steam to the evaporator. A major disadvantage of MVR system is the handling of large volume of vapour which requires a high capacity compressor. The running cost of compressor must be balanced with the steam economy of the evaporator.

Thermal Vapour Recompression (TVR) System

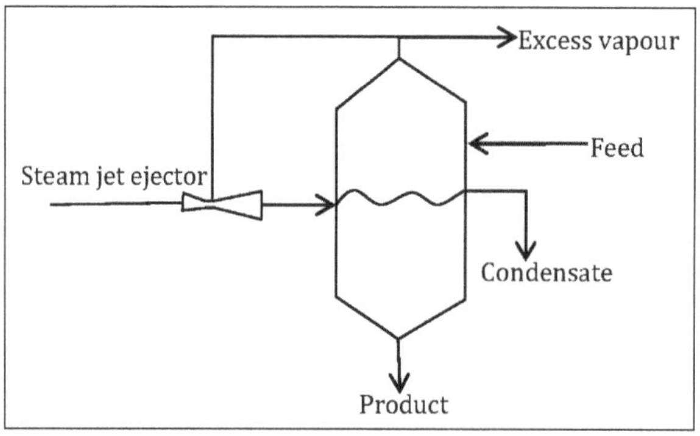

In order to avoid the difficulties in handling huge volume of vapour, thermal vapour recompression system is used. A steam jet ejector is the heart of the TVR system. Steam jet ejector is a device in which a high pressure steam passes through a narrow space, thus creates reduced pressure zone by increasing the velocity of steam. The design is similar to a nozzle with converging and diverging sides. A part of vapour from the evaporator is allowed to mix with the high pressure steam near the throat of the nozzle at right angle. Due to pressure difference between the evaporator and ejector, vapour is automatically dragged with the high jet steam which is then re-circulated to the evaporator. One advantage of TVR system is that it can handle high volume of vapour without any use of mechanical high friction moving parts. But, some precision should be maintained in the processing conditions like temperature and pressure for greater efficiency.

Boiling Point Elevation

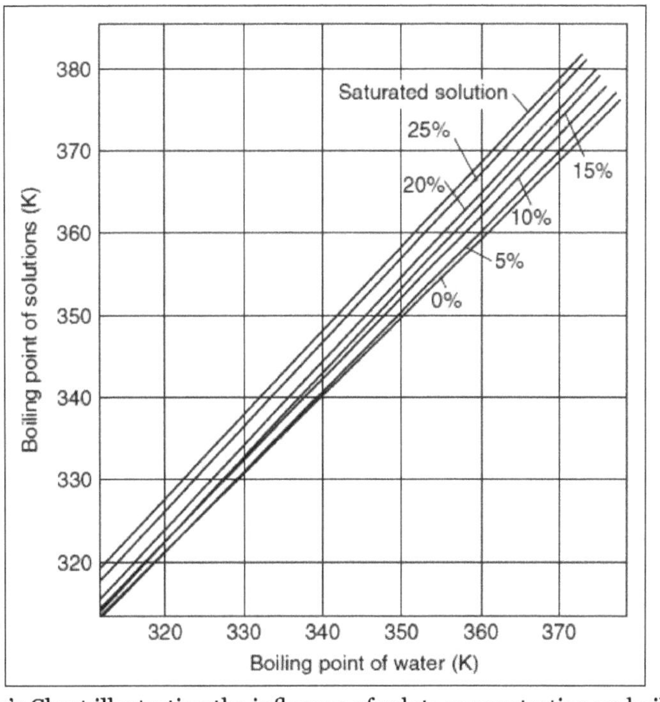

Figure: Duhring's Chart illustrating the influence of solute concentration on boiling point rise.

Boiling point elevation or boiling point rise of a solution is defined as the increase in boiling point over the boiling point of pure water at a particular pressure. Duhring's rule states that a linear relationship exists between the boiling point temperature of solution and the boiling point temperature of pure water at same pressure. This rule is valid in moderate range of temperature. Duhring's chart for sodium chloride solution at different concentration is shown in fig. From the figure it is concluded that at a particular boiling temperature of water, the boiling point of solution rises with the increased concentration. Say for example, at 0% concentration the salt behaves like water, the boiling points of both are same. Take the boiling temperature of 350 oK. When the concentration increases to 10%, the boiling point of solution becomes 354 °K. Hence, the boiling point rise is 4 °K at 10% salt concentration. Boiling point rise has a detrimental effect on the economy of evaporator as it reduces the temperature difference between the steam and the boiling solution, thus the heat transfer rate decreases.

Mechanical Separations

Mechanical separations can be divided into four groups - sedimentation, centrifugal separation, filtration and sieving.

In sedimentation, two immiscible liquids, or a liquid and a solid, differing in density, are separated by allowing them to come to equilibrium under the action of gravity, the heavier material falling with respect to the lighter. This may be a slow process. It is often speeded up by applying centrifugal forces to increase the rate of sedimentation; this is called centrifugal separation.

Filtration is the separation of solids from liquids, by causing the mixture to flow through fine pores which are small enough to stop the solid particles but large enough to allow the liquid to pass. Sieving, that is interposing a barrier through which the larger elements cannot pass, is often used for classification of solid particles.

Mechanical separation of particles from a fluid uses forces acting on these particles. The forces can be direct restraining forces such as in sieving and filtration, or indirect as in impingement filters. They can come from gravitational or centrifugal action, which can be thought of as negative restraining forces, moving the particles relative to the containing fluid. So the separating action depends on the character of the particle being separated and the forces on the particle which cause the separation. The important characteristics of the particles are size, shape and density; and of the fluid are viscosity and density. The reactions of the different components to the forces set up relative motion between the fluid and the particles, and between particles of different character. Under these relative motions, particles and fluids accumulate in different regions and can be gathered as in:

- The filter cake and the filtrate tank in the filter press.
- The discharge valve in the base of the cyclone and the air outlet at the top.
- The outlet streams of a centrifuge.
- On the various sized sieves of a sieve set.

In the mechanical separations studied, the forces considered are gravity, combinations of gravity with other forces, centrifugal forces, pressure forces in which the fluid is forced away from the particles, and finally total restraint of solid particles where normally the fluid is of little consequence. The velocities of particles moving in a fluid are important for several of these separations.

The Velocity of Particles Moving in a Fluid

Under a constant force, for example the force of gravity, particles in a liquid accelerate for a time and thereafter move at a uniform velocity. This maximum velocity which they reach is called their terminal velocity. The terminal velocity depends upon the size, density and shape of the particles, and upon the properties of the fluid.

When a particle moves steadily through a fluid, there are two principal forces acting upon it, the external force causing the motion and the drag force resisting motion which arises from frictional action of the fluid. The net external force on the moving particle is applied force less the reaction force exerted on the particle by the surrounding fluid, which is also subject to the applied force, so that,

$$F_s = V_a \left(\rho_p - \rho_f\right),$$

where F_s is the net external accelerating force on the particle, V is the volume of the particle, a is the acceleration which results from the external force, ρ_p is the density of the particle and ρ_f is the density of the fluid.

The drag force on the particle (F_d) is obtained by multiplying the velocity pressure of the flowing fluid by the projected area of the particle,

$$F_d = C\rho_{fv}{}^2 A / 2,$$

where C is the coefficient known as the drag coefficient, ρ_f is the density of the fluid, v is the velocity of the particle and A the projected area of the particle at right angles to the direction of the motion.

If these forces are acting on a spherical particle so that V = $\pi D^3/6$ and A = $\pi D^2/4$, where D is the diameter of the particle, then equating F_s and F_d, in which case the velocity v becomes the terminal velocity v_m, we have:

$$(\pi D^3 / 6) \times a(\rho_p - \rho_f) = C\rho_f v_m{}^2 \pi D^2 / 8.$$

It has been found, theoretically, that for the streamline motion of spheres, the coefficient of drag is given by the relationship:

$$C = 24/(Re) = 24\mu / Dv_m\rho_f.$$

Substituting this value for C and rearranging, we arrive at the equation for the terminal velocity,

$$v_m = D \;\; a(\rho_p - \rho_f)/18\mu.$$

This is the fundamental equation for movement of particles in fluids.

Sedimentation

Sedimentation uses gravitational forces to separate particulate material from fluid streams. The particles are usually solid, but they can be small liquid droplets, and the fluid can be either a liquid or a gas. Sedimentation is very often used in the food industry for separating dirt and debris from incoming raw material, crystals from their mother liquor and dust or product particles from air streams.

In sedimentation, particles are falling from rest under the force of gravity. Therefore in sedimentation, equation $v_m = D^2 \; a(\rho_p - \rho_f)/18\mu$ takes the familiar form of Stokes' Law:

$$v_m = D^2 \; g(\rho_p - \rho_f)/18\mu.$$

Note that equation $v_m = D^2 \; g(\rho_p - \rho_f)/18\mu$ is not dimensionless and so consistent units must be employed throughout. For example in the SI system, D in m, g in ms^{-2}, ρ in kgm-3 and μ in Nsm^{-2}, and then vm would be in ms^{-1}. Particle diameters are usually very small and are often measured in microns (micro-metres) = 10^{-6} m with the symbol μm.

Stoke's Law applies only in streamline flow and strictly only to spherical particles. In the case of spheres, the criterion for streamline flow is that (Re) ≤ 2, and many practical cases occur in the region of streamline flow, or at least where streamline flow is a reasonable approximation.

Example: Settling velocity of dust particles.

Calculate the settling velocity of dust particles of (a) 60μm and (b) 10μm diameter in air at 21°C and 100kPa pressure. Assume that the particles are spherical and of density 1280kgm³, and that the viscosity of air = 1.8 x 10^{-5} Ns m^{-2} and density of air = 1.2kgm^{-3}.

(a) For 60mm particle:

$$v_m = \frac{\left(60 \times 10^{-6}\right)^2 \times 9.81 \times (1280 - 1.2)}{\left(18 \times 1.8 \times 10^{-5}\right)}$$

$$= 0.14 ms^{-1}$$

(b) For 10μm particles since v_m is proportional to the squares of the diameters,

$$v_m = 0.14 \times (10/60)^2$$
$$= 3.9 \times 10^{-3} ms^{-1}$$

Checking the Reynolds number for the 60μm particles,

$$(Re) = (Dv\rho_f / \mu)$$
$$= \left(60 \times 10^{-6} \times 0.14 \times 1.2\right) / \left(1.8 \times 10^{-5}\right)$$
$$= 0.56$$

Stokes' Law applies only to cases in which settling is free, that is where the motion of one particle is unaffected by the motion of other particles. Where particles are in concentrated suspensions, an appreciable upward motion of the fluid accompanies the motion of particles downward. So the particles interfere with the flow patterns round one another as they fall. Stokes' Law predicts velocities proportional to the square of the particle diameters. In concentrated suspensions, it is found that all particles appear to settle at a uniform velocity once a sufficiently high level of concentration has been reached. Where the size range of the particles is not much greater than 10:1, all the particles tend to settle at the same rate. This rate lies between the rates that would be expected from Stokes' Law for the largest and for the smallest particles. In practical cases, in which Stokes Law or simple extensions of it cannot be applied, probably the only satisfactory method of obtaining settling rates is by experiment.

Gravitational Sedimentation of Particles in a Liquid

Solids will settle in a liquid whose density is less than their own. At low concentration, Stokes' Law will apply but in many practical instances the concentrations are too high. In a cylinder in which a uniform suspension is allowed to settle, various quite well defined zones appear as the settling proceeds. At the top is a zone of clear liquid. Below this is a zone of more or less constant composition, constant because of the uniform settling velocity of all sizes of particles. At the bottom of the cylinder is a zone of sediment with the larger particles further down. If the size range of the particles is wide, the zone of constant composition near the top will not occur and an extended zone of variable composition will replace it.

In a continuous thickener with settling proceeding as the material flows through, and in which clarified liquid is being taken from the top and sludge from the bottom, these same zones occur.

The minimum area necessary for a continuous thickener can be calculated by equating the rate of sedimentation in a particular zone to the counter flow velocity of the rising fluid. In this case we have:

$$v_u = (F - L)(dw/dt)/A\rho$$

where v_u is the upward velocity of the flow of the liquid, F is the mass ratio of liquid to solid in the feed, L is the mass ratio of liquid to solid in the underflow liquid, dw/dt is the mass rate of feed of the solids, ρ is the density of the liquid and A is the settling area in the tank.

If the settling velocity of the particles is v, then v_u =v and, therefore:

$$A = (F - L)(dw/dt)/v\rho$$

The same analysis applies to particles (droplets) of an immiscible liquid as to solid particles. Motion between particles and fluid is relative, and some particles may in fact rise.

Example: Separating of oil and water.

A continuous separating tank is to be designed to follow after a water washing plant for liquid oil. Estimate the necessary area for the tank if the oil, on leaving the washer, is in the form of globules 5.1×10^{-5}m diameter, the feed concentration is 4kg water to 1kg oil, and the leaving water is effectively oil free. The feed rate is $1000 kgh^{-1}$, the density of the oil is $894 kgm^{-3}$ and the temperature of the oil and of the water is 38oC. Assume Stokes' Law.

- Viscosity of water = $0.7 \times 10^{-3} Nsm^{-2}$
- Density of water at 38°C = 992 kgm^{-3}
- Diameter of globules = 5.1×10^{-5} m
- From equation $v_m = D^2 g(\rho_p - \rho_f)/18\mu$ $v_m = D^2 g(\rho_p - \rho_f)/18\mu$

  $V_m = (5.1 \times 10^{-5})\ 2 \times 9.81 \times (992 - 894)/(18 \times 0.7 \times 10^{-3})$

  $= 1.98 \times 10\text{-}4 ms^{-1}$

  $= 0.71\ mh^{-1}$

and since F = 4 and L = 0, and dw/dt = flow of minor component = 1000/5 = 200 kg h^{-1}, we have from equation $A = (F - L)(dw/dt)/v\rho$.

$A = (F - L)(dw/dt)/v\rho$

$A = 4 \times 200/(0.71 \times 1000)$

$= 1.1 m^2$

Sedimentation equipment for separation of solid particles from liquids by gravitational sedimentation is designed to provide sufficient time for the sedimentation to occur and to permit the overflow and the sediment to be removed without disturbing the separation. Continuous flow through the equipment is generally desired, so the flow velocities have to be low enough to avoid disturbing the sediment. Various shaped vessels are used, with a sufficient cross-section to keep the velocities down and fitted with slow speed, scraper conveyors and pumps to remove the settled solids. When vertical cylindrical tanks are used, the scrapers generally rotate about an axis in the centre of the tank and the overflow may be over a weir round the periphery of the tank, as shown diagrammatically in figure.

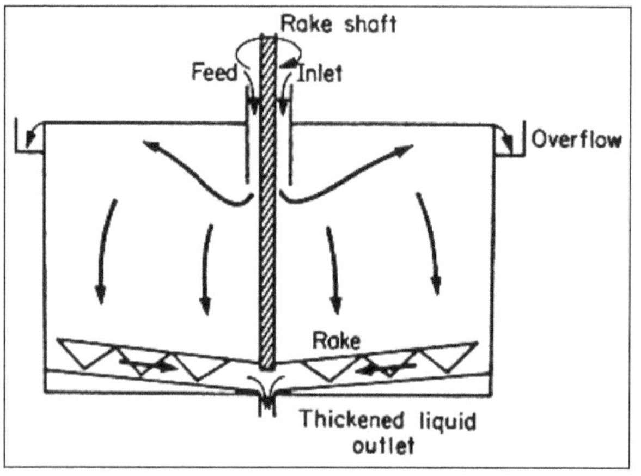

Figure: Continuous-sedimentation plant.

Flotation

In some cases, where it is not practicable to settle out fine particles, these can sometimes be floated to the surface by the use of air bubbles. This technique is known as flotation and it depends upon the relative tendency of air and water to adhere to the particle surface. The water at the particle surface must be displaced by air, after which the buoyancy of the air is sufficient to carry both the particle and the air bubble up through the liquid.

Because it depends for its action upon surface forces, and surface forces can be greatly changed by the presence of even minute traces of surface active agents, flotation may be promoted by the use of suitable additives. In some instances, the air bubbles remain round the solid particles and cause froths. These are produced in vessels fitted with mechanical agitators; the agitators whip up the air/liquid mixture and overflow the froth into collecting troughs.

The greatest application of froth flotation is in the concentration of minerals, but one use in the food industry is in the separation of small particles of fat from water. Dissolving the air in water under pressure provides the froth. On the pressure being suddenly released, the air comes out of solution in the form of fine bubbles which rise and carry the fat with them to surface scrapers.

Sedimentation of Particles in a Gas

In the food industry, an important application of sedimentation of solid particles occurs in spray dryers. In a spray dryer, the material to be dried is broken up into small droplets of about 100μm diameter and this fall through heated air, drying as they do so. The necessary area for the particles to settle can be calculated in the same way as for sedimentation. Two disadvantages arise from the slow rates of sedimentation: the large chamber areas required and also the long contact times between particles and the heated air which may lead to deterioration of heat sensitive products.

Settling under Combined Forces

It is sometimes convenient to combine more than one force to effect a mechanical separation. In consequence of the low velocities, especially of very small particles, obtained when gravity is the only external force acting on the system, it is well worthwhile to also employ centrifugal forces. Probably the most common application of this is the cyclone separator. Combined forces are also used in some powder classifiers such as the rotary mechanical classifier and in ring dryers.

Cyclones

Cyclones are often used for the removal from air streams of particles of about 10μm or more in diameter. They are also used for separating particles from liquids and for separating liquid droplets from gases. The cyclone is a settling chamber in the form of a vertical cylinder, so arranged that the particle-laden air spirals round the cylinder to create centrifugal forces which throw the particles to the outside walls. Added to the gravitational forces, the centrifugal action provides reasonably rapid settlement rates. The spiral path, through the cyclone, provides sufficient separation time. A cyclone is illustrated in figure.

Stokes' Law shows that the terminal velocity of the particles is related to the force acting. In a centrifugal separator, such as a cyclone, for a particle rotating round the periphery of the cyclone:

$$F_c = (mv^2)/r$$

where F_c is the centrifugal force acting on the particle, m is the mass of the particle, v is the tangential velocity of the particle and r is the radius of the cyclone.

This equation shows that the force on the particle increases as the radius decreases, for a fixed velocity. Thus, the most efficient cyclones for removing small particles are those of smallest diameter. The limitations on the smallness of the diameter are both the capital costs of small diameter cyclones to provide sufficient output, and also the pressure drops.

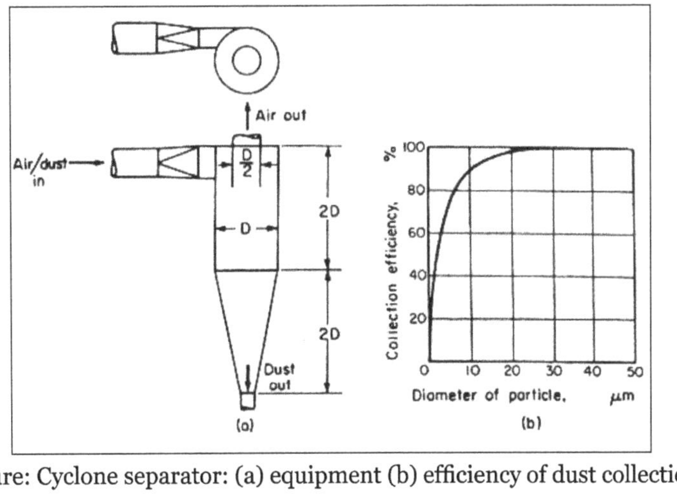

Figure: Cyclone separator: (a) equipment (b) efficiency of dust collection.

The optimum shape for a cyclone has been evolved mainly from experience and proportions similar to those indicated in Figure, have been found effective. The efficient operation of a cyclone depends very much on a smooth double helical flow being produced and anything which creates a flow disturbance or tends to make the flow depart from this pattern will have considerable and adverse effects upon efficiency. For example, it is important that the air enters tangentially at the top. Constricting baffles or lids should be avoided at the outlet for the air.

The efficiency of collection of dust in a cyclone is illustrated in figure. Because of the complex flow, the size cut of particles is not sharp and it can be seen that the percentage of entering particles which are retained in the cyclone falls off for particles below about 10μm diameter. Cyclones can be used for separating particles from liquids as well as from gases and also for separating liquid droplets from gases.

Impingement Separators

Other mechanical flow separators for particles in a gas use the principal of impingement in which deflector plates or rods, normal to the direction of flow, abruptly change the direction of flow. The gas recovers its direction of motion more rapidly than the particles because of its lower inertia. Suitably placed collectors can then be arranged to collect the particles as they are thrown out of the stream. This is the principle of operation of mesh and fibrous air filters. Various adaptations of impingement and settling separators can be adapted to remove particles from gases, but where the particle diameters fall below about 5μm, cloth filters and packed tubular filters are about the only satisfactory equipment.

Classifiers

Classification implies the sorting of particulate material into size ranges. Use can be made of the different rates of movement of particles of different sizes and densities suspended in a fluid and differentially affected by imposed forces such as gravity and centrifugal fields, by making suitable arrangements to collect the different fractions as they move to different regions.

Rotary mechanical classifiers, combining differential settling with centrifugal action to augment the force of gravity and to channel the size fractions so that they can be collected, have come into increasing use in flour milling. One result of this is that because of small differences in sizes, shapes and densities between starch and protein-rich material after crushing, the flour can be classified into protein-rich and starch-rich fractions. Rotary mechanical classifiers can be used for other large particle separation in gases.

Classification is also employed in direct air dryers, in which use is made of the density decrease of material on drying. Dry material can be sorted out as a product and wet material returned for further drying. One such dryer uses a scroll casing through which the mixed material is passed, the wet particles pass to the outside of the casing and are recycled while the material in the centre is removed as dry product.

Centrifugal Separations

The separation by sedimentation of two immiscible liquids, or of a liquid and a solid, depends on the effects of gravity on the components. Sometimes this separation may be very slow because the specific gravities of the components may not be very different, or because of forces holding the components in association, for example as occur in emulsions. Also, under circumstances when sedimentation does occur there may not be a clear demarcation between the components but rather a merging of the layers. For example, if whole milk is allowed to stand, the cream will rise to the top and there is eventually a clean separation between the cream and the skim milk. However, this takes a long time, of the order of one day, and so it is suitable, perhaps, for the farm kitchen but not for the factory. Much greater forces can be obtained by introducing centrifugal action, in a centrifuge. Gravity still acts and the net force is a combination of the centrifugal force with gravity as in the cyclone. Because in most industrial centrifuges, the centrifugal forces imposed are so much greater than gravity, the effects of gravity can usually be neglected in the analysis of the separation.

The centrifugal force on a particle that is constrained to rotate in a circular path is given by,

$$F_c = mr\omega^2$$

where F_c is the centrifugal force acting on the particle to maintain it in the circular path, r is the radius of the path, m is the mass of the particle, and ω(omega) is the angular velocity of the particle.

Or, since $\omega = v/r$, where v is the tangential velocity of the particle,

$$F_c = \left(mv^2\right)/r$$

Rotational speeds are normally expressed in revolutions per minute, so that equation $F_c = \left(mv^2\right)/r$ can also be written, as $\omega = 2\pi N/60$; as it has to be in s^{-1}, divide by 60.

$$F_c = mr\ (2\pi N/60)^2 = 0.011\ mrN^2$$

where N is the rotational speed in revolutions per minute.

If this is compared with the force of gravity (F_g) on the particle, which is $F_g = mg$, it can be seen that the centrifugal acceleration, equal to $0.011rN^2$, has replaced the gravitational acceleration, equal to g. The centrifugal force is often expressed for comparative purposes as so many "g".

Example: Centrifugal force in a centrifuge.

How many "g" can be obtained in a centrifuge which can spin a liquid at 2000 rev/min at a maximum radius of 10cm?

$$F_c = 0.011\ mrN^2$$
$$F_g = mg$$

$$F_c/F_g = \left(0.011rN^2\right)/g$$
$$= \left(0.011 \times 0.1 \times 2000^2\right)/9.81$$
$$= 450g.$$

The centrifugal force depends upon the radius and speed of rotation and upon the mass of the particle. If the radius and the speed of rotation are fixed, then the controlling factor is the weight of the particle so that the heavier the particle the greater is the centrifugal force acting on it. Consequently, if two liquids, one of which is twice as dense as the other, are placed in a bowl and the bowl is rotated about a vertical axis at high speed, the centrifugal force per unit volume will be twice as great for the heavier liquid as for the lighter. The heavy liquid will therefore move to occupy the annulus at the periphery of the bowl and it will displace the lighter liquid towards the centre. This is the principle of the centrifugal liquid separator, illustrated diagrammatically in figure.

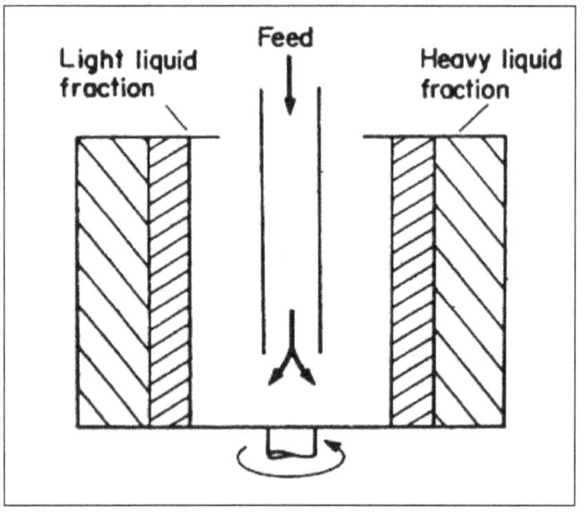

Figure: Liquid separation in a centrifuge.

Rate of Separation

The steady-state velocity of particles moving in a streamline flow under the action of an accelerating force is, from equation $v_m = D^2\ a(\rho_p - \rho_f)/18\mu$,

$$v_m = D^2\ a(\rho_p - \rho_f)/18\mu$$

If a streamline flow occurs in a centrifuge we can write, from equations $F_c = (mv^2)/r$ and $F_c = mr\ (2\pi N/60)^2 = 0.011\ mrN^2$, as a is the tangential acceleration:

$$F_c = ma$$

$$F_c / m = a = r(2\pi N/60)^2$$

so that,

$$v_m = D^2\ r(2\pi N/60)^2 (\rho_p - \rho_f)/18\mu$$

$$v_m = D^2 N^2\ r(\rho_p - \rho_f)/1640\mu$$

Example: Centrifugal separation of oil in water.

A dispersion of oil in water is to be separated using a centrifuge. Assume that the oil is dispersed in the form of spherical globules 5.1×10^{-5} m diameter; its density is 894 kgm^{-3}. If the centrifuge rotates at 1500 rev/mm and the effective radius at which the separation occurs is 3.8cm, calculate the velocity of the oil through the water. Take the density of water to be 1000kgm^{-3} and its viscosity to be 0.7×10^{-3} Nsm^{-2}. (The separation in this problem is the same as that in Example, in which the rate of settling under gravity was calculated).

From equation $v_m = D^2N^2\, r(\rho_p - \rho_f)/1640\mu$

$$v_m = (5.1 \times 10^{-5})^2 \times (1500)^2 \times 0.038 \times (1000 - 894)/(1.64 \times 10^3 \times 0.7 \times 10^{-3})$$
$$= 0.02 ms^{-1}$$

Checking that it is reasonable to assume Stokes' Law:

$$Re = (Dv\rho/\mu) = (5.1 \times 10^{-5} \times 0.02 \times 1000)/(0.7 \times 10^{-3})$$
$$= 1.5$$

so that the flow is streamline and it should obey Stokes' Law.

Liquid Separation

The separation of one component of a liquid/liquid mixture, where the liquids are immiscible but finely dispersed, as in an emulsion, is a common operation in the food industry. It is particularly common in the dairy industry in which the emulsion, milk, is separated by a centrifuge into skim milk and cream. It seems worthwhile, on this account, to examine the position of the two phases in the centrifuge as it operates. The milk is fed continuously into the machine, which is generally a bowl rotating about a vertical axis, and cream and skim milk come from the respective discharges. At some point within the bowl there must be a surface of separation between cream and the skim milk.

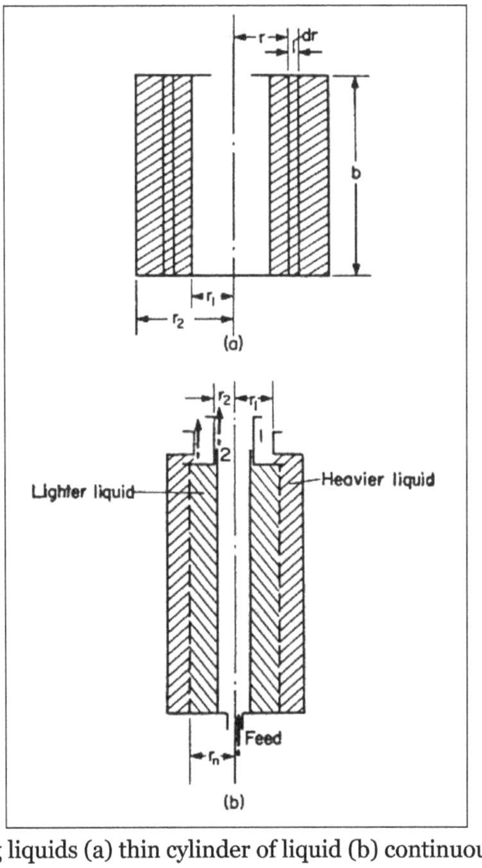

Figure: Centrifuging liquids (a) thin cylinder of liquid (b) continuous liquid centrifuge.

Consider a thin cylinder, of thickness d_r and height b as shown in figure: the differential centrifugal force across the thickness d_r is given by equation $F_c = mr\omega^2$:

$$dF_c = (dm)r\omega^2$$

where dF_c is the differential force across the cylinder wall, d_m is the mass of the differential cylinder, ω is the angular velocity of the cylinder and r is the radius of the cylinder. But,

$$dm = 2\omega rb\rho dr.$$

Where ρ is the density of the liquid and b is the height of the cylinder.

Therefore $dF_c = (2\pi rb\rho dr)r\omega^2$

The area over which the force dF_c acts is $2\pi rb$, so that:

$$dP = dF_c / 2\pi rb = \rho\omega^2\ rdr$$

Where dP is the differential pressure across the wall of the differential cylinder.

To find the differential pressure in a centrifuge, between radius r_1 and r_2, the equation for dP can be integrated, letting the pressure at radius r_1 be P_1 and that at r_2 be P_2, and so,

$$P_2 - P_1 = \rho\omega^2\ (r_2^2 - r_1^2)/2$$

Equation above shows the radial variation in pressure across the centrifuge.

Consider now figure, which represents the bowl of a vertical continuous liquid centrifuge. The feed enters the centrifuge near to the axis, the heavier liquid, A, discharges through the top opening 1 and the lighter liquid, B, through the opening 2. Let r_1 be the radius at the discharge pipe for the heavier liquid and r_2 that for the lighter liquid. At some other radius r_n, there will be a separation between the two phases, the heavier and the lighter. For the system to be in hydrostatic balance, the pressures of each component at radius r_n must be equal, so that applying equation $P_2 - P_1 = \rho\omega^2\ (r_2^2 - r_1^2)/2$ to find the pressures of each component at radius r_n, and equating these we have:

$$\rho_A\omega^2(r_n^2 - r_1^2)/2 = \rho_B\omega^2(r_n^2 - r_2^2)/2$$

$$\rho_A(r_n^2 - r_1^2) = \rho_B(r_n^2 - r_2^2)$$

$$r_n^2 = (\rho_A r_1^2 - \rho_B r_2^2)/(\rho_A - \rho_B)$$

where ρ_A is the density of the heavier liquid and ρB is the density of the lighter liquid.

Equation $r_n^2 = (\rho_A r_1^2 - \rho_B r_2^2)/(\rho_A - \rho_B)$ shows that as the discharge radius for the heavier liquid is made smaller, then the radius of the neutral zone must also decrease. When the neutral zone is nearer to the central axis, the lighter component is exposed only to a relatively small centrifugal force compared with the heavier liquid. This is applied where, as in the separation of cream from milk, as much cream as possible is to be removed and the neutral radius is therefore kept small. The feed to a centrifuge of this type should be as nearly as possible into the neutral zone so that it will enter with the least disturbance of the system. This relationship can, therefore, be used to place the feed inlet and the product outlets in the centrifuge to get maximum separation.

Example: Centrifugal separation of milk and cream.

If a cream separator has discharge radii of 5 cm and 7.5 cm and if the density of skim milk is 1032kgm^{-3} and that of cream is 915 kgm^{-3}, calculate the radius of the neutral zone so that the feed inlet can be designed.

For skim milk, r_1 = 0.075m, ρ_A = 1032 kgm^{-3}; for cream r_2 = 0.05m, ρ_B = 915 kgm^{-3}

From equation $r_n^{\,2} = (\rho_A r_1^{\,2} - \rho_B r_2^{\,2}) / (\rho_A - \rho_B)$

$$r_n^{\,2} = \left[1032 \times (0.075)^2 - 915 \times (0.05)^2\right] / (1032 - 915)$$

$$= 0.03\text{m}^2$$

$$r_n = 0.17 \text{ m}$$

$$= 17\text{cm}$$

Centrifuge Equipment

The simplest form of centrifuge consists of a bowl spinning about a vertical axis, as shown in figure. Liquids, or liquids and solids, are introduced into this and under centrifugal force the heavier liquid or particles pass to the outermost regions of the bowl, whilst the lighter components move towards the centre.

If the feed is all liquid, then suitable collection pipes can be arranged to allow separation of the heavier and the lighter components. Various arrangements are used to accomplish this collection effectively and with a minimum of disturbance to the flow pattern in the machine. To understand the function of these collection arrangements, it is very often helpful to think of the centrifuge action as analogous to gravity settling, with the various weirs and overflows acting in just the same way as in a settling tank even though the centrifugal forces are very much greater than gravity.

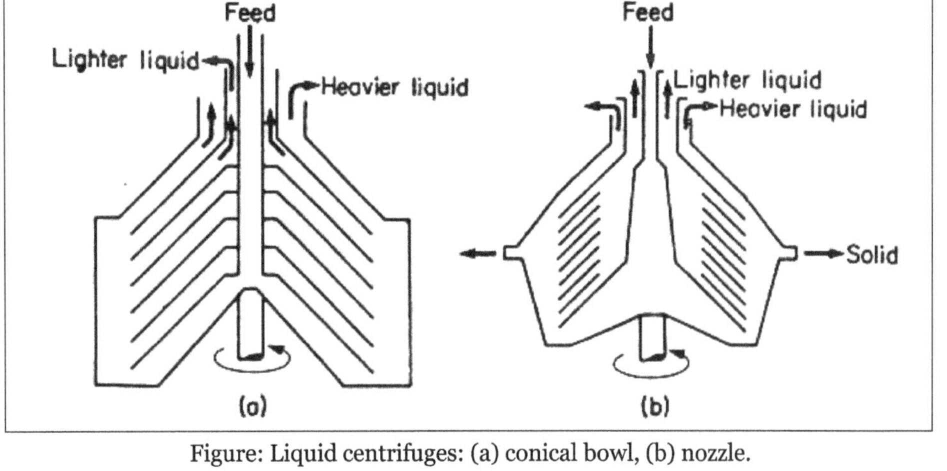

Figure: Liquid centrifuges: (a) conical bowl, (b) nozzle.

In liquid/liquid separation centrifuges, conical plates are arranged as illustrated in figure and these give smoother flow and better separation.

Whereas liquid phases can easily be removed from a centrifuge, solids present much more of a problem. In liquid/solid separation, stationary ploughs cannot be used as these create too much disturbance of the flow pattern on which the centrifuge depends for its separation. One method of handling solids is to provide nozzles on the circumference of the centrifuge bowl as illustrated in Figure. These nozzles may be opened at intervals to discharge accumulated solids together with some of the heavy liquid. Alternatively, the nozzles may be open continuously relying on their size and position to discharge the solids with as little as possible of the heavier liquid. These machines thus separate the feed into three streams, light liquid, heavy liquid and solids, the solids carrying with them some of the heavy liquid as well. Another method of handling solids from continuous feed is to employ telescoping action in the bowl, sections of the bowl moving over one another and conveying the solids that have accumulated towards the outlet, as illustrated in figure.

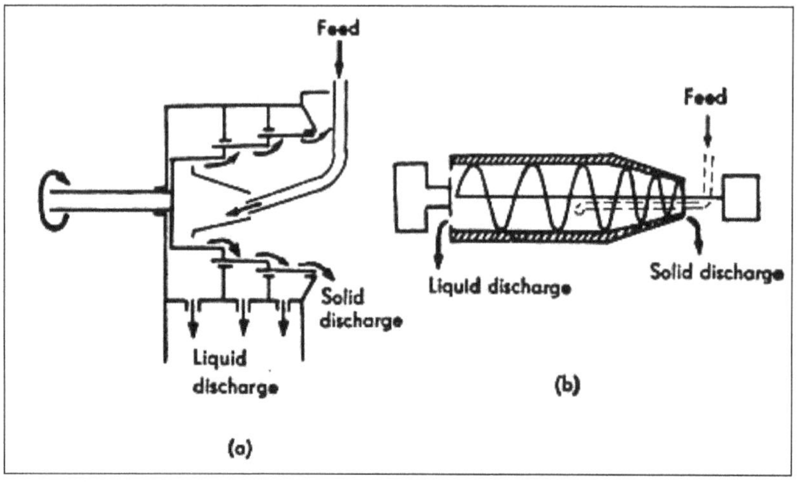

Figure: Liquid/solid centrifuges(a) telescoping bowl(b) horizontal bowl, scroll discharge.

The horizontal bowl centrifuge with scroll discharge, as illustrated in Figure can discharge continuously. In this machine, the horizontal collection scroll (or screw) rotates inside the conical-ended bowl of the machine and conveys the solids with it, whilst the liquid discharges over an overflow towards the centre of the machine and at the opposite end to the solid discharge. The essential feature of these machines is that the speed of the scroll, relative to the bowl, must not be great. For example, if the bowl speed is 2000rev/min, a suitable speed for the scroll might be 25rev/min relative to the bowl, which would mean a scroll speed of 2025 or 1975rev/min. The differential speeds are maintained by gearing between the driving shafts for the bowl and the scroll. These machines can continuously handle feeds with solid contents of up to 30%. A discussion of the action of centrifuges is given by Trowbridge and they are also treated in McCabe and Smith and Coulson and Richardson.

Filtration

In another class of mechanical separations, placing a screen in the flow through which they cannot pass imposes virtually total restraint on the particles above a given size. The fluid in this case is subject to a force that moves it past the retained particles. This is called filtration. The particles suspended in the fluid, which will not pass through the apertures, are retained and build up into what is called a filter cake. Sometimes it is the fluid; the filtrate that is the product, in other cases the filter cake.

The fine apertures necessary for filtration are provided by fabric filter cloths, by meshes and screens of plastics or metals, or by beds of solid particles. In some cases, a thin preliminary coat of cake, or of other fine particles, is put on the cloth prior to the main filtration process. This preliminary coating is put on in order to have sufficiently fine pores on the filter and it is known as a pre-coat.

The analysis of filtration is largely a question of studying the flow system. The fluid passes through the filter medium, which offers resistance to its passage, under the influence of a force which is the pressure differential across the filter. Thus, we can write the familiar equation:

Rate of filtration = driving force/resistance.

Resistance arises from the filter cloth, mesh, or bed, and to this is added the resistance of the filter cake as it accumulates. The filter-cake resistance is obtained by multiplying the specific resistance of the filter cake that is its resistance per unit thickness, by the thickness of the cake. The resistances of the filter material and pre-coat are combined into a single resistance called the filter resistance. It is convenient to express the filter resistance in terms of a fictitious thickness of filter cake. This thickness is multiplied by the specific resistance of the filter cake to give the filter resistance. Thus the overall equation giving the volumetric rate of flow d V/dt is:

$$dV / dt = (A \, \Delta P) / R$$

As the total resistance is proportional to the viscosity of the fluid, we can write:

$$R = \mu r (L_c + L)$$

where R is the resistance to flow through the filter, μ is the viscosity of the fluid, r is the specific resistance of the filter cake, Lc is the thickness of the filter cake and L is the fictitious equivalent thickness of the filter cloth and pre-coat, A is the filter area, and ΔP is the pressure drop across the filter.

If the rate of flow of the liquid and its solid content are known and assuming that all solids are retained on the filter, the thickness of the filter cake can be expressed by:

$$L_c = wV / A$$

where w is the fractional solid content per unit volume of liquid, V is the volume of fluid that has passed through the filter and A is the area of filter surface on which the cake forms.

The resistance can then be written,

$$R = \mu r \left[w(V / A) + L \right]$$

and the equation for flow through the filter, under the driving force of the pressure drop is then:

$$dV / dt = A \, \Delta P / \mu r \left[w(V / A) + L \right]$$

Equation above may be regarded as the fundamental equation for filtration. It expresses the rate of filtration in terms of quantities that can be measured, found from tables, or in some cases estimated. It can be used to predict the performance of large-scale filters on the basis of laboratory or pilot scale tests. Two applications of equation above are filtration at a constant flow rate and filtration under constant pressure.

Constant Rate Filtration

In the early stages of a filtration cycle, it frequently happens that the filter resistance is large relative to the resistance of the filter cake because the cake is thin. Under these circumstances, the resistance offered to the flow is virtually constant and so filtration proceeds at a more or less constant rate. Equation $dV/dt = A\,\Delta P/\mu r\left[w(V/A)+L\right]$ can then be integrated to give the quantity of liquid passed through the filter in a given time. The terms on the right-hand side of equation $dV/dt = A\,\Delta P/\mu r\left[w(V/A)+L\right]$ are constant so that integration is very simple:

$$\int dV/Adt = V/At = \Delta P/\left[\mu rw(V/A)+L\right]$$

Or $\Delta P = V/At \times \mu r\left[(wV/A)+L\right]$

From equation above the pressure drop required for any desired flow rate can be found. Also, if a series of runs is carried out under different pressures, the results can be used to determine the resistance of the filter cake.

Constant Pressure Filtration

Once the initial cake has been built up, and this is true of the greater part of many practical filtration operations, flow occurs under a constant-pressure differential. Under these conditions, the term ΔP in equation $dV/dt = A\,\Delta P/\mu r\left[w(V/A)+L\right]$ is constant and so,

$$\mu r\left[w(V/A)+L\right]dV = A\,\Delta P dt$$

and integration from V = 0 at t = 0, to V = V at t = t

$\mu r\left[w\left(V^2/2A\right)+LV\right] = A\,\Delta P t$ and rewriting this,

$$tA/V = (\mu rw/2\Delta P)\times(V/A)+\mu rL/\Delta P$$

$$t/(V/A) = (\mu rw/2\Delta P)\times(V/A)+\mu rL/\Delta P$$

Equation above is useful because it covers a situation that is frequently found in a practical filtration plant. It can be used to predict the performance of filtration plant on the basis of experimental results. If a test is carried out using constant pressure, collecting and measuring the filtrate at measured time intervals, a filtration graph can be plotted of t/(V/A) against (V/A) and from the statement of equation above it can be seen that this graph should be a straight line. The slope of this line will correspond to $\mu rw/2\Delta P$; the intercept on the t/(V/A) axis will give the value of $\mu rL/\Delta P$. Since, in general, μ, w, ΔP and A are known or can be measured, the values of the slope and intercept on this graph enable L and r to be calculated.

Example: Volume of filtrate from a filter press.

A filtration test was carried out, with particular product slurry, on a laboratory filter press under a constant pressure of 340 kPa and volumes of filtrate were collected as follows:

Filtrate volume (kg)	20	40	60	80
Time (min)	8	26	54.5	93

The area of the laboratory filter was 0.186 m^2. In a plant scale filter, it is desired to filter a slurry containing the same material, but at 50% greater concentration than that used for the test, and under a pressure of 270kPa. Estimate the quantity of filtrate that would pass through in 1 hour if the area of the filter is 9.3m^2.

From the experimental data:

V(kg)	20	40	60	80
t(s)	480	1560	3270	5580
V/A (l/m^2)	107.5	215	323	430
t/(V/A) ($sm^2\ kg^{-1}$)	4.47	7.26	10.12	12.98

These values of t/(V/A) are plotted against the corresponding values of V/A in Figure. From the graph, we find that the slope of the line is 0.0265, and the intercept 1.6.

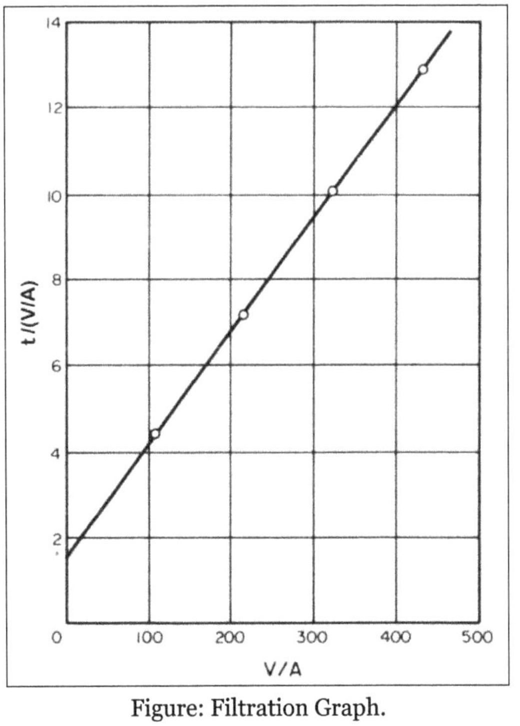

Figure: Filtration Graph.

Then substituting in equation $tA/V = (\mu rw/2\Delta P)\times(V/A)+\mu rL/\Delta P$ we have,

$$t/(V/A)=0.0265(V/A)+1.6.$$

To fit the desired conditions for the plant filter, the constants in this equation will have to be modified. If all of the factors in equation $tA/V = (\mu rw/2\Delta P)\times(V/A)+\mu rL/\Delta P$ except those which are varied in the problem are combined into constants, K and K', we can write,

$$t/(V/A)=(w/\Delta P)K \times (V/A)+ K'/\Delta P(a)$$

In the laboratory experiment $w = w_1$, and $\Delta P = \Delta P_1$,

$$K=(0.0265\Delta P_1/w_1) \text{ and } K'=1.6\Delta P_1.$$

For the new plant condition, $w = w_2$ and $P = P_2$, so that, substituting in the equation (a) above, we then have for the plant filter, under the given conditions:

$$t/(V/A)=(0.0265\,\Delta P_1/w_1)(w_2/\Delta P_2)(V/A)+(1.6\Delta P_1)(1/\Delta P_2)$$

and since from these conditions,

$$\Delta P_1/\Delta P_2=340/270$$

and $w_2/w_1=150/100$,

$$t/(V/A)=0.0265(340/270)(150/100)(V/A)+1.6(340/270)$$
$$=0.05(V/A)+2.0$$
$$t = 0.5(V/A)^2+2.0(V/A).$$

To find the volume that passes the filter in 1 h which is 3600s, that is to find V for t = 3600.

$$3600= 0.05(V/A)^2 + 2.0(V/A)$$

and solving this quadratic equation, we find that V/A = 250kgm^{-2}

and so the slurry passing through 9.3 m^2 in 1 h would be:

= 250 × 9.3

= 2325kg.

Filter-cake Compressibility

With some filter cakes, the specific resistance varies with the pressure drop across it. This is because the cake becomes denser under the higher pressure and so provides fewer and smaller passages for flow. The effect is spoken of as the compressibility of the cake. Soft and flocculent materials provide highly compressible filter cakes, whereas hard granular materials, such as sugar and salt crystals, are little affected by pressure. To allow for cake compressibility the empirical relationship has been proposed:

$$r= r'\Delta(P)^s$$

where r is the specific resistance of the cake under pressure P, ΔP is the pressure drop across the filter, r' is the specific resistance of the cake under a pressure drop of 1atm and s is a constant for the material, called its compressibility.

This expression for r can be inserted into the filtration equations, such as equation $tA/V = (\mu rw/2\Delta P)\times(V/A)+\mu rL/\Delta P$, and values for r' and s can be determined by carrying out experimental runs under various pressures.

Filtration Equipment

The basic requirements for filtration equipment are:

- Mechanical support for the filter medium.
- Flow accesses to and from the filter medium.
- Provision for removing excess filter cake.

In some instances, washing of the filter cake to remove traces of the solution may be necessary. Pressure can be provided on the upstream side of the filter, or a vacuum can be drawn downstream, or both can be used to drive the wash fluid through.

Plate and Frame Filter Press

In the plate and frame filter press, a cloth or mesh is spread out over plates which support the cloth along ridges but at the same time leave a free area, as large as possible, below the cloth for flow of the filtrate. This is illustrated in figure. The plates with their filter cloths may be horizontal, but they are more usually hung vertically with a number of plates operated in parallel to give sufficient area.

Filter cake builds up on the upstream side of the cloth, that is the side away from the plate. In the early stages of the filtration cycle, the pressure drop across the cloth is small and filtration proceeds at more or less a constant rate. As the cake increases, the process becomes more and more a constant pressure one and this is the case throughout most of the cycle. When the available space between successive frames is filled with cake, the press has to be dismantled and the cake scraped off and cleaned, after which a further cycle can be initiated.

The plate and frame filter press is cheap but it is difficult to mechanize to any great extent. Variants of the plate and frame press have been developed which allow easier discharging of the filter cake. For example, the plates, which may be rectangular or circular, are supported on a central hollow shaft for the filtrate and the whole assembly enclosed in a pressure tank containing the slurry.

Filtration can be done under pressure or vacuum. The advantage of vacuum filtration is that the pressure drop can be maintained whilst the cake is still under atmospheric pressure and so can be removed easily. The disadvantages are the greater costs of maintaining a given pressure drop by applying a vacuum and the limitation on the vacuum to about 80kPa maximum. In pressure filtration, the pressure driving force is limited only by the economics of attaining the pressure and by the mechanical strength of the equipment.

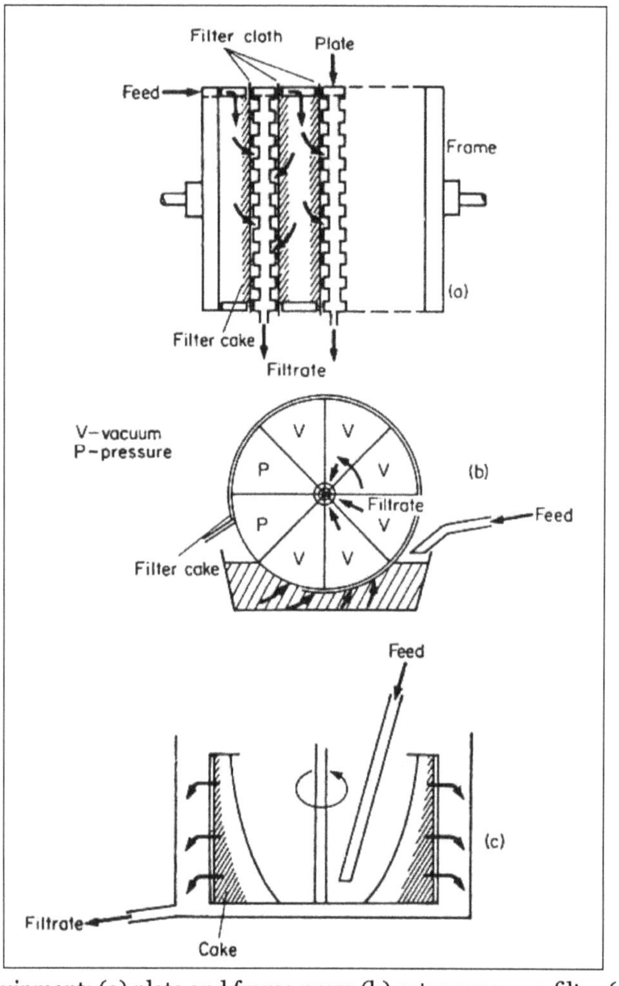

Figure: Filtration equipment: (a) plate and frame press (b) rotary vacuum filter (c) centrifugal filter.

Rotary Filters

In rotary filters, the flow passes through a rotating cylindrical cloth from which the filter cake can be continuously scraped. Either pressure or vacuum can provide the driving force, but a particularly useful form is the rotary vacuum filter. In this, the cloth is supported on the periphery of a horizontal cylindrical drum that dips into a bath of the slurry. Vacuum is drawn in those segments of the drum surface on which the cake is building up. A suitable bearing applies the vacuum at the stage where the actual filtration commences and breaks the vacuum at the stage where the cake is being scraped off after filtration. Filtrate is removed through trunnion bearings. Rotary vacuum filters are expensive, but they do provide a considerable degree of mechanization and convenience. A rotary vacuum filter is illustrated diagrammatically in Figure.

Centrifugal Filters

Centrifugal force is used to provide the driving force in some filters. These machines are really centrifuges fitted with a perforated bowl that may also have filter cloth on it. Liquid is fed into the interior of the bowl and under the centrifugal forces, it passes out through the filter material. This is illustrated in figure.

Air Filters

Filters are used quite extensively to remove suspended dust or particles from air streams. The air or gas moves through a fabric and the dust is left behind. These filters are particularly useful for the removal of fine particles. One type of bag filter consists of a number of vertical cylindrical cloth bags 15-30cm in diameter, the air passing through the bags in parallel. Air bearing the dust enters the bags, usually at the bottom and the air passes out through the cloth. A familiar Example of a bag filter for dust is to be found in the domestic vacuum cleaner. Some designs of bag filters provide for the mechanical removal of the accumulated dust. For removal of particles less than 5μm diameter in modern air sterilization units, paper filters and packed tubular filters are used. These cover the range of sizes of bacterial cells and spores.

Sieving

In the final separation operation in this group, restraint is imposed on some of the particles by mechanical screens that prevent their passage. This is done successively, using increasingly smaller screens, to give a series of particles classified into size ranges. The fluid, usually air, can effectively be ignored in this operation which is called sieving. The material is shaken or agitated above a mesh or cloth screen; particles of smaller size than the mesh openings can pass through under the force of gravity. Rates of throughput of sieves are dependent upon a number of factors:

- Nature and shape of the particles.
- Frequency and amplitude of the shaking.
- Methods used to prevent sticking or bridging of particles in the apertures of the sieve.
- Tension and physical nature of the sieve material.

Standard sieve sizes have evolved, covering a range from 25mm aperture down to about 0.6mm aperture. The mesh was originally the number of apertures per inch. A logical base for a sieve series would be that each sieve size have some fixed relation to the next larger and to the next smaller. A convenient ratio is 2:1 and this has been chosen for the standard series of sieves in use in the United States, the Tyler sieve series. The mesh numbers are expressed in terms of the numbers of opening to the inch (= 2.54 cm). By suitable choice of sizes for the wire from which the sieves are woven, the ratio of opening sizes has been kept approximately constant in moving from one sieve to the next. Actually, the ratio of 2:1 is rather large so that the normal series progresses in the ratio of $\sqrt{2}$:1 and if still closer ratios are required intermediate sieves are available to make the ratio between adjacent sieves in the complete set $\sqrt[4]{2}$: 1. The standard British series of sieves has been based on the available standard wire sizes, so that, although apertures are generally of the same order as the Tyler series, aperture ratios are not constant. In the SI system, apertures are measured in mm.

In order to get reproducible results in accurate sieving work, it is necessary to standardize the procedure. The analysis reports either the percentage of material that is retained on each sieve, or the cumulative percentage of the material larger than a given sieve size.

The results of a sieve analysis can be presented in various forms, perhaps the best being the cumulative analysis giving, as a function of the sieve aperture (D), the weight fraction of the powder F(D) which passes through that and larger sieves, irrespective of what happens on the smaller ones. That is the cumulative fraction sums all particles smaller than the particular sieve of interest.

Thus F = F(D),

$$dF / dD = F'(D)$$

where F'(D) is the derivative of F(D) with respect to D.

So $\int dF = \int F'(D)\, dD$

and so integrating between D_1 and D_2 gives the cumulative fraction between two sizes D_2 (larger) and D_1 which is also that fraction passing through sieve of aperture D_2 and caught on that of aperture D_1. The F'(D) graph gives a particle size distribution analysis.

Example: Sieve analysis.

Given the following sieve analysis:

Sieve size mm	% Retained
1.00	0
0.50	11
0.25	49
0.125	28
0.063	8
Through 0.063	4

plot a cumulative sieve analysis and estimate the weight fraction of particles of sizes between 0.300 and 0.350 mm and 0.350 and 0.400 mm.

From the above table:

Less than aperture	(mm)	0.063	0.125	0.250	0.500	1.00
	(μm)	63	125	250	500	1000
Percentage (cumulative)		4	12	40	89	100

This has been plotted on figure and the graph F(D) has been smoothed. From this the graph of F'(D) has been plotted, working from the slope of F(D), to give the particle size distribution.

To find the fraction between the specified sizes, equation $\int dF = \int F'(D)\, dD$ indicates that this will be given directly by the fraction, that the area under the F'(D) graph and between the sizes of interest, is to the total area under the F'(D) curve. Counting squares, on figure, gives: between 300μm (0.300 mm) and 350μm (0.350mm) as 13% and 350μm (0.350mm) and 400μm (0.400mm) as 9%.

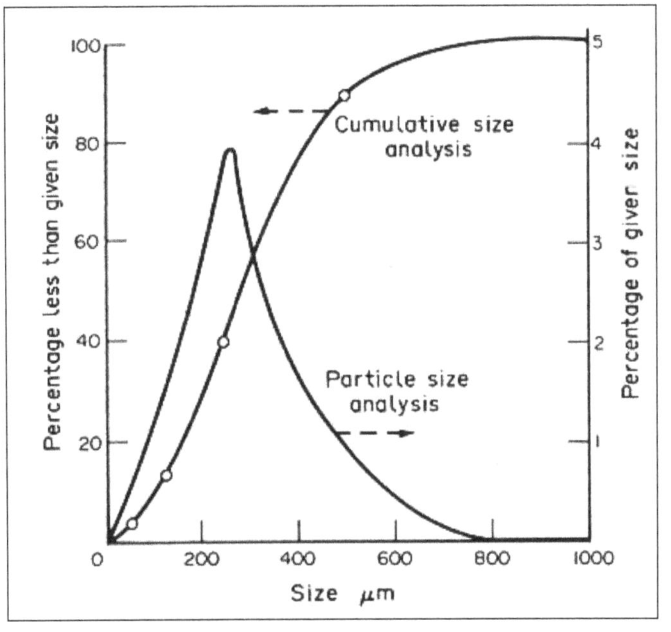

Figure: Particle-size analysis.

For industrial sieving, it is seldom worthwhile to continue until equilibrium is reached. In effect, a sieving efficiency term is introduced, as a proportion only of the particles smaller than a given size actually get through. The sieves of a series are often mounted one above the other, and a mechanical shaker used.

Sieve analysis for particle size determination should be treated with some caution especially for particles deviating radically from spherical shape, and it needs to be supplemented with microscopical examination of the powders. The size distribution of powders can be useful to estimate parameters of technological importance such as the surface area available for a reaction, the ease of dispersion in water of a dried milk powder, or the performance characteristics of a spray dryer or a separating cyclone.

Industrial sieves include rotary screens, which are horizontal cylinders either perforated or covered with a screen, into which the material is fed. The smaller particles pass through as they tumble around in the rotating screens. Other industrial sieves are vibrating screens, generally vibrated by an eccentric weight; and multi-deck screens on which the particles fall through from one screen to another, of decreasing apertures, until they reach one which is too fine for them to pass. With vibrating screens, the frequency and amplitude of the vibrations can significantly affect the separation achieved. Screen capacities are usually rated in terms of the quantity passed through per unit area in unit time. Particles that can conveniently be screened industrially range from 50μm diameter, upwards.

Continuous vibrating sieves used in the flour-milling industry employ a sieve of increasing apertures as particles progress along the length of the screen. So the finer fraction at any stage is being removed as the flour particles move along. The shaking action of the sieve provides the necessary motion to make the particles fall through and also conveys the oversize particles on to the next section. Below the sieves, in some cases, air classification may be used to remove bran.

References

- The-Relationship-Between-Heat-Transfer-and-Cooking: webstaurantstore.com, Retrieved 01, August 2020
- Heat-transfer-applications: org.nz, Retrieved 24, May 2020

Chapter 3

Techniques of Food Preservation

Some of the important food preservation techniques are curing, cooling, freezing, boiling, heating, pickling, sugaring, canning, etc. Curing is the process of removing moisture from the food by adding salt, sugar or nitrate and facilitating osmosis. The topics elaborated in this chapter will help in gaining a better perspective about these food preservation techniques.

Curing

Curing is the addition to meats of some combination of salt, sugar, nitrite and/or nitrate for the purposes of preservation, flavor and color. Some publications distinguish the use of salt alone as salting, corning or salt curingand reserve the word curing for the use of salt with nitrates/nitrites. The cure ingredients can be rubbed on to the food surface, mixed into foods dry (dry curing), or dissolved in water (brine, wet, or pickle curing). In the latter processes, the food is submerged in the brine until completely covered. With large cuts of meat, brine may also be injected into the muscle. The term pickle in curing has been used to mean any brine solution or a brine cure solution that has sugar added.

Salting/Corning

Salt inhibits microbial growth by plasmolysis. In other words, water is drawn out of the microbial cell by osmosis due to the higher concentration of salt outside the cell. A cell loses water until it reaches a state first where it cannot grow and cannot survive any longer. The concentration of salt outside of a microorganism needed to inhibit growth by plasmolysis depends on the genus and species of the microorganism. The growth of some bacteria is inhibited by salt concentrations as low as 3%, e.g., Salmonella, whereas other types are able to survive in much higher salt concentrations, e.g., up to 20% salt for Staphylococcus or up to 12% salt for Listeria monocytogenes. Fortunately the growth of many undesirable organisms normally found in cured meat and poultry products is inhibited at relatively low concentrations of salt.

Salting can be accomplished by adding salt dry or in brine to meats. Dry salting, also called corning originated in Anglo-Saxon cultures. Meat was dry-cured with coarse "corns" or pellets of salt. Corned beef of Irish fame is made from a beef brisket, although any cut of meat can be corned. Salt brine curing involves the creation of brine containing salt, water and other ingredients such as sugar, erythorbate, or nitrites. Age-old tradition was to add salt to the brine until it floated an egg. Today, however, it is preferred to use a hydrometer or to carefully mix measured ingredients from a reliable recipe. Once mixed and placed into a suitable container, the food is submerged in the salt brine. Brine curing usually produces an end product that is less salty compared to dry curing. Injection of brine into the meat can also speed the curing process.

Nitrate/Nitrite Curing

Most salt cures do not contain sufficient levels of salt to preserve meats at room temperature and Clostridium botulinum spores can survive. In the early 1800's it was realized that saltpeter ($NaNO_3$ or KNO_3) present in some impure curing salt mixtures would result in pink colored meat rather than the typical gray color attained with a plain salt cure. This nitrate/nitrite in the curing process was found to inhibit growth of Clostridium. Recent evidence indicates that they may also inhibit E. coli, Salmonella, and Campylobacter if in sufficient quantities.

Several published studies indicated that N-nitrosoamines were considered carcinogenic in animals. For this reason, nitrate is prohibited in bacon and the nitrite concentration is limited in other cured meats. In other cured foods, there is insufficient scientific evidence for N-nitrosamine formation and a link to cancer.

Cure Mixtures

For the home food preserver, measuring small batches of cure for nitrites or nitrates would require an analytical scale that few consumers have access to. Therefore, some manufacturers sell premixed salt and nitrate/nitrite curing mixes for easy home use. Caution is needed when using pure saltpeter instead of commercially prepared mixes, since accidental substitution of saltpeter for table salt in recipes can result in lethal toxic levels.

Some examples of commercially prepared cures include:

Prague Powder, Insta Cure or Modern Cure

This cure contains sodium nitrite (6.25%) mixed with salt (93.75%). Consumers are recommended to use 1 oz. for every 25 lb. of meat or one level teaspoon of cure for 5 lb. of meat.

Prague Powder

This mix is used for dry cured meats that require long (weeks to months) cures. It contains 1 oz. of sodium nitrite and 0.64 oz. of sodium nitrate. It is recommended that this cure be combined with each 1 lb. of salt and for products that do not require cooking, smoking, or refrigeration. This cure, which contains sodium nitrate, acts like a time-release cure, slowly breaking down into sodium nitrite, then into nitric oxide. The manufacturer recommends using 1 oz. of cure for 25 lbs. of meat or one level teaspoon of cure for 5 lbs. of meat.

Mixes

Many individual manufacturers and commercial sausage makers produce curing mixtures, often combining sugar and spices with the salt and nitrite/nitrates. It is important that consumers follow manufacturer directions carefully.

Saltpeter, Sodium or Potassium Nitrate

Commercially, nitrate is no longer allowed for use in curing of smoked and cooked meats, non-smoked and cooked meats, or sausages (US FDA 1999). However, nitrate is still allowed in small amounts in the making of dry cured uncooked products. Home food preservers should avoid the direct use of this chemical and opt for the mixtures described.

Combination Curing

Some current recipes for curing have vinegar, citrus juice, or alcohol as ingredients for flavor. Addition of these chemicals in sufficient quantities can contribute to the preservation of the food being cured.

Flavor of Cured Meats

Besides preservation, the process of curing introduces both a desired flavor and color. Cured meat flavor is thought to be a composite result of the flavors of the curing agents and those developed by bacterial and enzymatic action.

Salt

Because of the amount of salt used in most curing processes, the salt flavor is the most predominant.

Sugar

Sugar is a minor part of the composite flavor, with bacon being an exception. Because of the tremendous amount of salt used, sugar serves to reduce the harshness of the salt in cured meat and enhance the sweetness of the product (ie. Sweet Lebanon Bologna). Sugar also serves as a nutrient source for the flavor-producing bacteria of meat during long curing processes.

Spices and Flavor Enhancers

Spices add characteristic flavors to the meats. Recent studies have suggested that some spices can have added preservative effects (Doyle 1999). However, the quantities of spice needed to achieve these effects may be well beyond the reasonable quantities of use.

Nitrates/Nitrites

Nitrites and nitrate conversion to nitrite provide the characteristic cured flavor and color.

Fermentation

The tangy flavor observed in dry fermented sausages, such as pepperoni, is the result of bacterial fermentation or the addition of chemicals such as glucono-δ(*delta*)-lactone.

Smoking

The process of smoking gives the product the characteristic smoky flavor that can be varied slightly with cure recipes and types of smoke used.

Color of Cured Meats

A high concentration of salt promotes the formation of an unattractive gray color within some meat. Nitrate when used for some dry-cured, non-cooked meats is reduced to nitrite then to nitric oxide, which reacts with myoglobin (muscle pigment) to produce the red or pink cured color. If nitrite is used as the curing agent, there is no need for the nitrate reduction step, and the development of the cure color is much more rapid.

Generation of Nitric Oxide (NO):

$$NaNO_3 \xrightarrow{(1)} NaNO_2 \xrightarrow{(2)} HONO \xrightarrow{(3)} NO$$

1. Sodium nitrate is reduced to sodium nitrite by microorganisms such as Micrococcus spp. present on meats.
2. Sodium nitrite is reduced to nitrous acid in the presence of an acidic environment (e.g., by fermentation or by addition of glucono-δ(delta)-lactone).
3. Nitrous acid forms nitric oxide. Nitric oxide reacts with myoglobin (meat pigments) to form a red color.

The time required for a cured color to develop may be shortened with the use of cure accelerators, e.g., ascorbic acid, erythorbic acid, or their derivatives. Cure accelerators tend to speed up chemical conversion of nitric acid to nitric oxide. They also serve as oxygen scavengers, which slow the fading of the cured meat color in the presence of sunlight and oxygen. Some studies have indicated that cure accelerators have antimicrobial properties, especially for the newly emerging pathogens like *E. coli*O157:H7 and *Listeria monocytogenes* (Doyle 1999). Since cure accelerators are rarely used in home curing, this information needs further review or research to determine what benefits home curing would have by using certain cure accelerators.

Ham

Ham is cured pork from the hind leg of the hog. Picnic shoulder or picnic ham is made from the front leg of the hog. Ham varieties may or may not be smoked and are available in many regional and ethnic styles. Curing solutions for hams typically contain salt, sodium nitrate, sugar, and seasonings. Dry-cured ham includes country ham and proscuitto. The dry cure mixture is rubbed onto the pork surface and the meat is cured (at or below 40°F) from weeks to a year or more. During this aging process, the moisture is reduced by 18-25%, making these hams safe at room temperature. Brine-cured ham includes culatello and Irish Hams. Usually the fresh meat is both injected with brine and submerged into the brine to allow the cure to reach all of the meat.

Bacon

Bacon is cured or smoked hog meat from the pig belly. Bacon produced at home, is typically dry-cured with salt, nitrites, sugar, and spices for a week or longer. Because of concern over N-nitrosamines, the use of nitrates for bacon curing is not allowed commercially. Home preparations, such as Morton Smoked-flavored sugar cure, contain nitrates and are recommended by the manufacturer for the use in bacon curing. Some ethnic bacon (Canadian bacon and Irish bacon) is made from leaner cuts. Pancetta is Italian bacon that is not smoked. Salt pork is salted pork belly fat.

Beef

The most well known cured beef product is corned beef made from the beef brisket. Pastrami is smoked corned beef.

Poultry

Any variety of poultry can be cured and smoked. Curing and smoking imparts a unique, delicate flavor and pink color to poultry meat. As with other meats, curing and smoking increases the refrigerated storage life of poultry. When preparing smoked poultry products, most consumers use mild cures (relatively low salt) to maintain the poultry flavor.

Fish

Any fish can be salted and smoked. Some varieties of fish make for better tasting products than others. Commercially, nitrite curing is only allowed for sable, salmon, shad, chub, and tuna in the U.S. Other species were never included in the Code of Federal Regulations simply because industry members did not respond to initial inquiries about GRAS (Generally Regarded as Safe) practices (Ken Hilderbrand, personal communication).

Sausage

Sausage can be made from any meat source, and is typically ground. Sausage can be uncured and unsmoked, but for the purposes of this document, we consider only cured and/or smoked sausage. Usually cure ingredients (salt, nitrates/nitrites, and spices) are mixed with the ground meat and stuffed into casings (animal intestines or collagen). The product is then cured for a short time (e.g. overnight for bologna) at refrigerated temperatures. It may or may not be smoked, dried, or fermented.

Game

Venison, bear, elk, wild boar, wild turkey, rabbit and other game animals can be successfully cured/smoked.

Food Safety Concerns

Concern for food safety has arisen over:

1) The public's desire for variety and healthfulness that leads them to both non-traditional foods and non-traditional processes that may lack research into their safety.

2) The emergence of new foodborne diseases that challenge the safety of traditional food preservation methods. Bacteria, yeasts and molds find meat a suitable substrate for growth, resulting in meat quality and safety deterioration. Foodborne diseases are mostly of bacterial origin and meat has been implicated in roughly one third of the foodborne outbreaks in North America. The pathogenic microorganisms representing the greatest risk with meat

and poultry borne diseases are *Salmonella* spp., *Campylobacter* spp, verotoxigenic *Escherichia coli, Listeria monocytogenes* and *Toxoplasma gondii.* Consumers and home food preservers should be warned that microorganisms are ubiquitous in the environment and that pathogens may survive traditional and non-traditional food preservation techniques if they are improperly processed.

Non-traditional Foods and Non-traditional Processes

Today, consumers demand foods that are minimally processed, as "natural" as possible, and yet are convenient to use. Complicating these factors is a consumer preference toward cured and smoked foods that are processed with lower salt, lower nitrate and higher moisture levels. These parameters have a tremendous impact on the safety of a given cured/smoked food or process. Preferences for low fat and low sugar have less impact on the safety, but these factors can change the traditional curing and smoking process. It will be difficult to completely eliminate the use of nitrite, as there is no known substitute for it as a curing agent for meat. Nonetheless, the demand for fewer chemicals added to foods has put pressure on the industry and the scientific community to seek new alternatives.

In-home vacuum packaging machines have become popular in recent years. It is important to realize that in-home vacuum packaging is not a substitution for cooking or any form of food preservation, e.g., refrigeration, freezing, or curing. In-home vacuum packaging can reduce the quality deterioration of foods catalyzed by oxygen, such as rancidity. Many food spoilage and food poisoning organisms require oxygen for growth and would also be inhibited by this process. However, the most deadly food poisoning organism, *Clostridium botulinum* requires a low oxygen atmosphere and therefore, vacuum packaging favors its growth. In cured meats, careful attention must be paid to proper use of nitrates/nitrites that inhibit *Clostridium botulinum* prior to use of in-home vacuum packagers. To further reduce the risk of botulism after vacuum packaging, properly refrigerate the cured/smoked meats. Under normal processing, freezing of salt-cured meats is not recommended, due to oxidative rancidity that affects the quality and flavor of the product.

Emergence of New Foodborne Diseases

More than 200 known diseases are transmitted through food .The causes include viruses, bacteria, and parasites. Many of the pathogens causing foodborne illness were not recognized 20 years ago. Major emerging pathogens include *Campylobacter jejuni, Salmonella, Listeria monocytogenes,* and *Escherichia coli* O157:H7. Many emerging foodborne diseases can cause chronic and serious health problems.

Food Poisoning Organisms

Microorganisms are ubiquitous in foods. Some can be present and harmless. Others can be present and produce chemicals that alter the acceptability of the food, hence food spoilage. Lastly, microorganisms can be present where they themselves or the products they produce can cause food poisoning.

Botulism

The majority (65%) of botulism cases are a result of inadequate home food processing or preservation. Botulism results from ingestion of a toxin produced by the bacterium *C. botulinum*. This bacterium requires a moist, oxygen-free environment, low acidity (pH greater than 4.6) and temperatures in the danger zone (38-140°F) to grow and produce toxin. *C. botulinum* forms heat resistant spores that can become dangerous if allowed to germinate, grow, and produce toxin. Sufficient heat can be used to inactivate the toxin (180°F for 4 min., Kendall 1999). *C. botulinum* thrives in moist foods that are low in salt (less than 10%), particularly when they are stored at temperatures above 38°F. These organisms will not grow in an aerobic environment, but other aerobic organisms in a closed system can rapidly convert an aerobic environment to an anaerobic environment by using the oxygen for their own growth, permitting growth of *C. botulinum*.

Clostridium Perfringens

Spores of some strains of *Clostridium perfringens* are so heat resistant that they survive boiling for four or more hours. Furthermore, cooking drives off oxygen, kills competitive organisms, and heat-shocks the spores, all of which promote germination to vegetative or growing cells. Once the spores have germinated, a warm, moist, protein-rich environment with little or no oxygen is necessary for growth. If such conditions exist (i.e., incorrectly holding meats at warm room temperature for smoking), sufficient numbers of vegetative cells may be produced to cause illness upon ingestion of the contaminated meat product.

Listeria Monocytogenes

L. monocytogenes has been found in fermented raw-meat sausages, raw and cooked poultry, raw meats (all types), and raw and smoked fish. Its ability to grow at temperatures as low as 3°C, permits multiplication in refrigerated foods. The organism grows in the pH range of 5.0 to 9.5 and is resistant to freezing. It is salt tolerant and relatively resistant to drying, but easily destroyed by heat. (It grows between 34 - 113°F).

E. coli O157:H7

Ground beef is the food most associated with *E. coli* O157:H7 outbreaks, but smoked and cured foods also have been implicated, including dry-cured salami, game meat, and homemade venison jerky. Studies have shown that *E. coli* O157:H7 can survive the typical dry fermentation processing conditions; *E. coli*O157:H7's tolerance of acidic conditions has also been reported in the processing of other foods such as apple cider and mayonnaise. These findings led to significant changes in the food industry and in the manufacturing of dry fermented sausage in the U.S. In August 1995, USDA/FSIS recommended using a heat process (145°F for 4 minutes) to inhibit *E. coli* O157:H7 growth in sausage.

Trichinosis

Details on trichinosis can be found in a publication by the National Pork Producers Council (Gamble) and on trichinosis statistics in the USA (CDC 1988). Trichinosis is an infestation of trichinae, or *Trichinella spiralis* or other *Trichinella* spp. The parasites invade the muscles causing severe pain and edema. It can be avoided by ensuring that cooked pork or certain wild game meat reaches an internal temperature of 150°F or more. Freezing the pork according to the following chart also can kill trichinae:

Freezer Temperature	Group 1 Days	Group 2 Days
5°F	20	30
-10°F	10	20
-20°F	6	12
Group 1 comprises product in separate pieces not exceeding 6" in thickness or arranged on separate racks with the layers not exceeding 6" in depth. Group 2 comprises product in pieces, layers or within containers the thickness of which exceeds 6" but not 27".		

Freezing Pork to Kill Trichinae.

Although the incidence of trichinosis has decreased markedly from 300 to 400 cases annually in the 1940's to less than 90 cases per year in the early 1980's, this disease remains a problem in the United States. According to USDA recommendations, *T. spiralis* in pork is rendered non-viable if held at 5°F, a temperature achievable in noncommercial freezers, for 20 days. However, meat from wild game, such as polar bear or walrus meat that has been infected with *T. spiralis*, remains infective even after 24 months of storage at 0°F. The difference in susceptibility may be caused by different strains of *T. spiralis* found in domestic versus wild animals. Adequate cooking (170°F. internally), well above the thermal death point of the organism (137°F), remains the best safeguard against trichinosis in game meats.

Staphylococcus Aureus

Staphylococcus is more salt-tolerant than most other bacteria. It is naturally present on human skin. Some species of *Staphylococcus* produce toxins that cause food poisoning. So, handling of cured meats with unwashed hands, followed by holding the food at warm temperatures (>40°F), can result in bacterial growth and toxin formation. While temperatures of 120°F can kill the bacterium itself, its toxin is heat resistant; therefore, it is important to keep the *Staphylococcus* organism from growing in foods. Use proper food handling practices to avoid contact with potentially contaminated surfaces and materials. Keep food either hot (above 140°F) or cold (below 40°F) during serving time, and as quickly as possible, refrigerate or freeze leftovers and foods to be served later. *Staphylococcus aureus* is destroyed by cooking and other thermal processing, but can be reintroduced via mishandling; the bacteria can then produce a toxin that is not destroyed by further cooking. Dry curing may or may not destroy *S. aureus*, but the high salt content on the exterior of dry cured meats inhibits these bacteria. When the dry cured meat is sliced, the moist, lower salt interior will permit staphylococcal multiplication.

Salmonella

Salmonella outbreaks have been recorded for raw meats, poultry, and fish and beef jerky. Salmonella bacteria thrive at temperatures between 40-140°F. They are readily destroyed by cooking to 165°F and do not grow at refrigerator or freezer temperatures. They do survive refrigeration and freezing, however, and will begin to grow again once warmed to room temperature.

Campylobacter

Raw chicken is a primary source of this organism, which grows best in a reduced oxygen environment. It is easily killed by heat (120°F), is inhibited by acid, salt and drying, and will not multiply at temperatures below 85°F. *Campylobacter* is the leading bacterial cause of diarrhea in the U.S.

Vibrio

Infections with this organism have been associated with the consumption of raw, improperly cooked, or cooked and recontaminated fish and shellfish. A correlation exists between the probability of infection and warmer months of the year. Improper refrigeration of seafood contaminated with this organism will allow its proliferation, increasing the possibility of infection. People with liver disease are particularly at risk for infection caused by undercooked seafood containing *V. vulnificus*.

Parasites (Other than Trichinella)

Anisakis simplex parasites are known to occur frequently in the flesh of cod, haddock, fluke, pacific salmon, herring, flounder, and monkfish. However, only 10 reported cases annually in the U.S. are attributed to them. *Diphyllobothrium latum* and *Nanophyetus*spp. parasites are known to occur frequently in the flesh of fish. Foodborne illnesses attributed to them are few in number. Sufficient cooking of foods would destroy the parasites.

In the Great Lakes region of the U. S., the Broad Fish Tapeworm has resulted in food poisoning outbreaks related to pickled pike. The larvae pass through small fish until they hatch as small worms in larger fish. If consumed at this stage by humans the worms can grow in the intestines. Sufficient cooking of foods would destroy the parasites.

Viruses

Shellfish are the food most often implicated foods in outbreaks of viruses such as Norwalk and Hepatitis A. Ingestion of raw or insufficiently steamed clams and oysters poses a high risk for infection with viruses. Sufficient cooking of foods would destroy the viruses.

Inhibition of Pathogens in Cured Meats

Salt and nitrates or nitrites are the primary chemicals that are responsible for the inhibition of pathogen growth when curing meats. Adding to that, pH and temperature (below 40°F or above 140°F), these factors can act in concert to prohibit the growth of pathogens in these foods. Table indicates some extreme parameters for growth of pathogens.

Organism	min. pH	max. % salt	min. temp.	oxygen req.
Campylobacter	4.9	2	86°F	MA1
Clostridium	4.7	10	38°F	AN2
E. coli	3.6	8	33°F	FA3
Listeria	4.8	12	32°F	FA
Salmonella	4.0	8	41°F	FA
Staphylococcus	4.0	20	41°F	FA
Vibrio	3.6	10	41°F	FA
1MA=microaerophilic; requires limited levels of oxygen; 2AN=anaerobic, requires the absence of oxygen; and 3FA=facultative anaerobic, can grow either with or without oxygen.				

Critical Parameters for growth of some Pathogens.

Cured/Smoked Food Poisoning

Ham

Trichinella, *Staphylococcus*, and molds are the microorganisms most associated with ham. All ham should be processed to specifically kill trichinae. *Staphylococcus aureus*, which is salt tolerant, can, survive the high salt levels of the ham surface. Once the ham is sliced, *S. aureus* can grow on the interior tissues where there is a lower salt concentration. Therefore, the USDA-FSIS recommends that all sliced ham be refrigerated. Molds can grow on the ham surface, especially on country-cured hams. The USDA-FSIS recommends that you wash the ham free of the mold with a stiff vegetable brush and that consumption of the ham is safe. We were unable to find any studies of aflatoxin formation with molds associated with hams.

Bacon

Like other cured products, *Listeria monocytogenes* has been responsible for a number of recalls of ready-to-eat bacon, e.g., State of Ohio Department of Agriculture Recall Announcement (ODA/ODH) 99 05a. Packages stored at room temperature sampled positive for the pathogen.

Beef

Pastrami made in a small Idaho commercial firm tested positive for *Listeria monocytogenes* in July 2000. No reports of food poisonings were recorded, but the products were recalled.Corned beef samples also tested positive for *Listeria monocytogenes* from a Michigan commercial firm. Corned beef was cooked and temperature abused at a deli in Ohio resulting in an outbreak of *C. perfringens* food poisoning.

Poultry

Much of the reports of food poisoning and recalls of poultry products for have been with commercial ready to eat products, such as chicken or turkey lunchmeats.

Fish

Listeria monocytogenes has been found in commercial samples of cold smoked fish leading to product recalls in New York (Cold smoked sea bass FDA Recall No.F-313-1) and Seattle, WA (Cold smoked salmon FDA Recall #F-265-1). These recalls demonstrate that even with HACCP and careful plant sanitation, commercial processors have contamination incidences in their cold smoked fish processes. In New York, fish sausage was recalled because laboratory analysis found pH (acidity), salt and water activity levels in the product were such that they could potentially permit *Clostridium botulinum* to develop and produce the toxin.

Sausage

Recent concern about the safety of sausages has been in the semi-dry fermented sausages, such as summer sausage. *E. coli* O157:H7 has been found to survive the acidity of these products. Some commercial, ready-to-eat sausages and luncheon meats have been implicated in *Listeria monocytogenes* growth and outbreaks. Additional concerns with trichinae may occur in any pork sausage.

Game

Precaution should be used since venison, bear, elk, wild boar, wild turkey, rabbit and other game animals are usually field dressed in unknown sanitary conditions or kept from immediate refrigeration. Two areas of special interest should be noted: (1) *E. coli*O157:H7 outbreaks in game sausage and jerky, and (2) Trichinosis in game meats from northern U.S. areas. Several outbreaks of *E. coli*O157:H7 have occurred in venison jerky.

T. nativa is an Alaskan, Canadian, and Arctic strain of *Trichinella* that is freeze-resistant. Unlike pork, freezing arctic meat will not kill larval cysts. Wild game, e.g., bear or walrus meat, is safe once the entire piece is completely cooked. USDA recommends attaining an internal temperature of at least 170°F. Since cooking may be uneven, microwaving of game meats is not recommended.

Cured/Smoked Food Spoilage

Not all microbial growth leads to food poisoning. Indeed, many organisms simply spoil cured and smoked foods making them unpalatable. Keep in mind that it is a general rule that if conditions exist to allow growth of spoilage organisms, these same conditions can allow for the growth of food poisoning organisms. Good judgment should prevail.

Lactic Acid Bacteria

Lactic acid bacteria are frequent spoilage organisms on cured/smoked meats. They are tolerant of some of the conditions in the curing/smoking process or are contaminates after processing. They grow slowly, but eventually spoil the food by producing organic acids.

Mold and Cured Meats

Moldy cured or smoked meat is a controversial topic. Very often country hams will have a moldy surface. Currently the USDA FSIS recommends cleaning the mold and soaking the ham in water to refresh it is a safe procedure. Other suggestions are to wash the ham in acetic acid (acetic acid Avinegar@ 10% in water).

Greening of Cured/Smoked Meats

Lactobacillus viridescens, or similar bacteria that produce hydrogen peroxide may cause greening in meats. The H_2O_2 reacts with myoglobin to produce a green sheen pigment. The meat, while less appealing, is not dangerous to consume.

Slime Producers

Some *Micrococcus* spp. and other bacteria are capable of producing slime on the surface of hams, bacon, and sausages.

Gas Producers

Some organisms can produce gas pockets inside cured or smoked meats.

Rancid Flavors in Home Cured Pork

Salt increases oxidation during long cures and can lead to a rancid flavor. Prolonged frozen storage may also contribute to oxidation leading to rancid flavors. Many consumers prefer these flavors. For those that do not, shorter curing and aging times should be considered.

Cooling

Cooling can be defined as a processing technique that is used to reduce the temperature of the food from one processing temperature to another or to a required storage temperature. Chilling is a processing technique in which the temperature of a food is reduced and kept at a temperature between –1°C and 8°C. The objective of cooling and chilling is to reduce the rate of biochemical and microbiological changes in foods, in order to extend the shelf-life of fresh and processed foods, or to maintain a certain temperature in a food process, e.g. in the fermentation and treatment of beer. Cooling is also used to promote a change of state of aggregation, e.g. crystallisation. In the wine industry, cooling (chilling) is applied to clarify the must before fermentation. The objective of cold stabilisation is to obtain the precipitation of tartrates (in wines) or fatty acids (in spirits) before bottling.

Applications

Cooling is a process step in many food production processes. Chilling for food preservation is widely applied for a lot of perishable foods. The main application of cold stabilisation in the food industry is in the wine and spirit sector.

The supply of chilled foods to consumers requires a sophisticated distribution system, involving chilled stores, refrigerated transport and chilled retail display cabinets. Chilled foods can be grouped into three categories according to the storage temperature (Hendley, B. (1985) Market for chilled foods. Food Process 52, 29-33) and the forth is applied in wine making:

- -1°C to + 1°C (fresh fish, meats, sausages and ground meats, smoked meats and fish);

- 0°C to + 5°C (pasteurised canned meat, milk and milk products, prepared salads, baked goods, pizzas, unbaked dough and pastry);
- 0°C to + 8°C (fully cooked meats and fish pies, cooked or uncooked cured meats, butter, margarine, cheese and soft fruits);
- 8°C to 12°C in the wine industry. The must is kept at this temperature between 6 and 24 hours.

Techniques, Methods and Equipments

Cooling of liquid foods is commonly carried out by passing the product through a heat exchanger (cooler) or by cooling the vessels. The cooling medium in the cooler can be groundwater, water recirculating over a cooling tower, or water (eventually mixed with agents like glycol) which is recirculated via a mechanical refrigeration system (ice-water). Cooling and chilling of solid foods is carried out by contacting the food with cold air, or directly with a refrigerant like liquid carbon dioxide or liquid nitrogen. The equipment used for freezing can also be used for cooling and chilling.

Some typical applications are given below:

a) Cooling of sugar

Sugar to be stored in silos must be dedusted and cooled to the storage temperature. This is done in a sugar cooler, which is a device in which warm and dried sugar is intensively aerated by cold filtered external air to cool the sugar to the storage temperature, approximately 20–30°C. The most common systems in use are coolers (typically drum or fluidised-bed coolers) with chilling systems with countercurrent or cross-current phase flow.

b) Cryogenic cooling

In cryogenic cooling the food is in direct contact with the refrigerant, which may be solid or liquid carbon dioxide, or liquid nitrogen. As the refrigerant evaporates or sublimates it removes heat from the food, thereby causing rapid cooling. Both liquid nitrogen and carbon dioxide refrigerants are colourless, odourless and inert.

c) Cold stabilisation

Cold stabilisation is a technique of chilling wines before bottling to cause the precipitation of harmless tartrate crystals.

For spirits, this technique consists of bringing the spirit to a temperature of between -1°C and -7°C, depending on the operators, and possibly performing a stabulation (storing at low temperature) in a tank at constant temperature for between 24 and 48 hours. A cold filtration (around -1°C) allows the fatty acid esters to be retained.

For wines, three techniques can be employed: stabilisation by batch and stabulation. This is the oldest technique and consists of bringing the wine to a temperature below zero close to the freezing point, then stabulating in an isothermal tank during a period of 5 to 8 days.

But currently the most widely-used techniques are: continuous stabilisation, where the stabulation tank is replaced by a cylindro-conical crystalliser and an agitator, in which the wine will remain for only between 30 and 90 minutes, stabilisation by crystal seeding consisting of refrigerating at between -1° and -2°C, and seeding at 4 g/l of tartaric crystals with agitation over 2 to 4 hours, and later storage in tank, and decantation after 12 to 48 hours. There can be many variations on these basic schemes.

Freezing

Freezing is one of the oldest and most widely used methods of food preservation, which allows preservation of taste, texture, and nutritional value in foods better than any other method. The freezing process is a combination of the beneficial effects of low temperatures at which microorganisms cannot grow, chemical reactions are reduced, and cellular metabolic reactions are delayed.

The Importance of Freezing as a Preservation Method

Freezing preservation retains the quality of agricultural products over long storage periods. As a method of long-term preservation for fruits and vegetables, freezing is generally regarded as superior to canning and dehydration, with respect to retention in sensory attributes and nutritive properties. The safety and nutrition quality of frozen products are emphasized when high quality raw materials are used, good manufacturing practices are employed in the preservation process, and the products are kept in accordance with specified temperatures.

The Need for Freezing and Frozen Storage

Freezing has been successfully employed for the long-term preservation of many foods, providing a significantly extended shelf life. The process involves lowering the product temperature generally to -18 °C or below. The physical state of food material is changed when energy is removed by cooling below freezing temperature. The extreme cold simply retards the growth of microorganisms and slows down the chemical changes that affect quality or cause food to spoil.

Competing with new technologies of minimal processing of foods, industrial freezing is the most satisfactory method for preserving quality during long storage periods. When compared in terms of energy use, cost, and product quality, freezing requires the shortest processing time. Any other conventional method of preservation focused on fruits and vegetables, including dehydration and canning, requires less energy when compared with energy consumption in the freezing process and storage. However, when the overall cost is estimated, freezing costs can be kept as low (or lower) as any other method of food preservation.

Current Status of Frozen Food Industry in U.S. and other Countries

The frozen food market is one of the largest and most dynamic sectors of the food industry. In spite of considerable competition between the frozen food industry and other sectors, extensive quantities of frozen foods are being consumed all over the world. The industry has recently grown to a value of over US$ 75 billion in the U.S. and Europe combined. This number has reached US$ 27.3 billion in 2001 for total retail sales of frozen foods in the U.S. alone (AFFI, 2003). In Europe, based on U.S. currency, frozen food consumption also reached 11.1 million tons in 13 countries in the year 2000. represents the division of frozen food industry in terms of annual sales in 2001.

Advantages of Freezing Technology in Developing Countries

Developed countries, mostly the U.S., dominate the international trade of fruits and vegetables. The U.S. is ranked number one as both importer and exporter, accounting for the highest percent of fresh produce in world trade. However, many developing countries still lead in the export of fresh exotic fruits and vegetables to developed countries.

For developing countries, the application of freezing preservation is favorable with several main considerations. From a technical point of view, the freezing process is one of the most convenient and easiest of food preservation methods, compared with other commercial preservation techniques. The availability of different types of equipment for several different food products results in a flexible process in which degradation of initial food quality is minimal with proper application procedures. As mentioned earlier, the high capital investment of the freezing industry usually plays an important role in terms of economic feasibility of the process in developing countries. As for cost distribution, the freezing process and storage in terms of energy consumption constitute approximately 10 percent of the total cost. Depending on the government regulations, especially in developing countries, energy cost for producers can be subsidized by means of lowering the unit price or reducing the tax percentage in order to enhance production. Therefore, in determining the economical convenience of the process, the cost related to energy consumption (according to energy tariffs) should be considered.

Food items	Sales US$ (million)	% Change vs. 2000
Total Frozen Food Sales	26 600	6.1
Baked Goods	1 400	9.0
Breakfast Foods	1 050	4.1
Novelties	1 900	10.5
Ice Cream	4 500	5.7
Frozen Dessert/Fruit/Toppings	786	5.4
Juices/Drinks	827	-9.7
Vegetables	2 900	4.3

Frozen food industry in terms of annual sales in 2001.

Increasing Consumer Demand in Developing Countries due to Modernization

The proportion of fresh food preserved by freezing is highly related to the degree of economic development in a society. As countries become wealthier, their demand for high-valued commodities increases, primarily due to the effect of income on the consumption of high-valued commodities in developing countries. The commodities preserved by freezing are usually the most perishable ones, which also have the highest price. Therefore, the demand for these commodities is less in developing areas. Besides, the need for adequate technology for freezing process is the major drawback of developing countries in competing with industrialized countries. The frozen food industry requires accompanying developments and facilities for transporting, storing, and marketing their products from the processing plant to the consumer. Thus, a large amount of capital investment is needed for these types of facilities. For developing countries, especially in rural or semi-rural areas, the frozen food industry has therefore not been developed significantly compared to other countries.

In recent years, due to the changing consumer profile, the frozen food industry has changed significantly. The major trend in consumer behavior documented over the last half century has been the increase in the number of working women and the decline in the family size. These two factors resulted in a reduction in time spent preparing food. The entry of more women into the workforce also led to improvements in kitchen appliances and increased the variability of ready-to-eat or frozen foods available in the market. Besides, the increased usage of microwave ovens, affecting food habits in general and the frozen food market in particular, as well as allowing rapid preparation of meals and greater flexibility in meal preparation. The frozen food industry is now only limited by imagination, an output of which increases continuously to supply the increasing demand for frozen products and variability.

Country	1999	2000	2001	2002
Argentina	n.a.	0.075	0.069	n.a.
Belgium	0.056	0.048	n.a.	n.a.
Bolivia	n.a.	0.062	0.069	n.a.
Chile	n.a.	0.052	0.056	n.a.
Chinese Taipei (Taiwan)	0.058	0.061	0.056	n.a.
Colombia	n.a.	0.052	0.042	n.a.
Costa Rica	n.a.	0.068	0.076	n.a.
Cuba	n.a.	0.080	0.078	n.a.
Ecuador	n.a.	0.036	0.061	n.a.
El Salvador	n.a.	0.111	0.110	n.a.
Finland	0.046	0.039	0.038	0.043
Germany	0.057	0.041	0.044	n.a.
Greece	0.050	0.042	0.043	0.046
Guyana	n.a.	0.082	0.080	n.a.
Hungary	0.055	0.049	0.051	0.060
India	0.081	0.080	n.a.	n.a.
Ireland	0.057	0.049	0.060	0.075
Italy	0.086	0.089	n.a.	n.a.

Korea (Korea, South)	0.056	0.062	0.057	n.a.
Mexico	0.042	0.051	0.053	n.a.
Netherlands	0.061	0.057	0.059	n.a.
New Zealand	0.030	0.030	0.028	0.033
Nicaragua	n.a.	0.117	0.115	n.a.
Paraguay	n.a.	0.032	0.036	n.a.
Peru	n.a.	0.056	0.057	n.a.
Poland	0.037	0.037	0.045	0.049
Portugal	0.078	0.067	0.066	0.068
Russia	0.012	0.011	n.a.	n.a.
South Africa	0.017	0.017	0.013	n.a.
Spain	0.049	0.043	0.041	n.a.
Switzerland	0.090	0.069	0.069	0.073
Turkey	0.079	0.080	0.079	0.094
United Kingdom	0.064	0.055	0.048	n.a.
United States 2	0.044	0.046	0.050	0.048
Uruguay	n.a.	0.064	0.070	n.a.

Unit electricity prices for industry1 (U.S. Dollars per Kilowatt-hour).

n.a. = Not Available.

- Energy end-use prices including taxes converted using exchange rates
- Electricity prices in the United States, including income taxes, environmental charges, and other charges.

Market Share of Frozen Fruits and Vegetables

Today in modern society, frozen fruits and vegetables constitute a large and important food group among other frozen food products. The historical development of commercial freezing systems designed for special food commodities helped shape the frozen food market. Technological innovations as early as 1869 led to the commercial development and marketing of some frozen foods. Early products saw limited distribution through retail establishments due to insufficient supply of mechanical refrigeration. Retail distribution of frozen foods gained importance with the development of commercially frozen vegetables in 1929.

The frozen vegetable industry mostly grew after the development of scientific methods for blanching and processing in the 1940s. Only after the achievement of success in stopping enzymatic degradation, did frozen vegetables gain a strong retail and institutional appeal. Today, market studies indicate that considering overall consumption of frozen foods, frozen vegetables constitute a very significant proportion of world frozen-food categories (excluding ice cream) in Austria, Denmark, Finland, France, Germany, Italy, Netherlands, Norway, Sweden, Switzerland, UK, and the USA. The division of frozen vegetables in terms of annual sales in 2001.

Commercialization history of frozen fruits is older than frozen vegetables. The commercial freezing of small fruits and berries began in the eastern part of the U.S. in about 1905. The main advantage of freezing preservation of fruits is the extended usage of frozen fruits during off-season.

Additionally, frozen fruits can be transported to remote markets that could not be accessed with fresh fruit. Also, freezing preservation makes year-round further processing of fruit products possible, such as jams, juice, and syrups from frozen whole fruit, slices, or pulps. In summary, the preservation of fruits by freezing has clearly become one the most important preservation methods.

Future Trends in Freezing Technology

The frozen food industry is highly based in modern science and technology. Starting with the first historical development in freezing preservation of foods, today, a combination of several factors influences the commercialization and usage of freezing technology. The future growth of frozen foods will mostly be affected by economical and technological factors. Growth in population, personal incomes, relative cost of other forms of foods, changes in tastes and preferences, and technological advances in freezing methods are some of the factors concerned with the future of freezing technology.

Population growth and increasing demand for food has generated the need for commercial production of food commodities in large-scale operations. Thus, availability of proper equipment suitable for continuous processing would be valuable for freezing preservation methods. In addition depending on personal incomes, relative cost of frozen products is one of the most important of economical factors. Producing the highest quality at the lowest cost possible is highly dependent on the technology used. As a result, developments in freezing technology in recent years have mostly been characterized by the improvements in mechanical handling and process control to increase freezing rate and reduce cost.

Today an increasing demand for frozen foods already exits and further expansion of the industry is primarily dependent on the ability of food processors to develop higher qualities in both process techniques and products. Improvements can only be achieved by focusing on new technologies and investigating poorly understood factors that influence the quality of frozen food products. Improvements in new and convenient forms of foods, as well as more information on relative cost and nutritive values of frozen foods, will contribute toward continued growth of the industry.

Vegetables	Sales US$ (million)	% Change vs. 2000
Broccoli	184	4.4
Com/Corn on the Cob	312	3.5
Green Beans	115	6.0
Mixed Vegetables	450	7.2
Peas	207	3.9
Potatoes	1 070	4.4

Frozen vegetables in terms of annual sales in 2001.

General Recommendations on the Freezing Process

Freezing is a widely used method of food preservation based on several advantages in terms of retention of food quality and ease of process. Beginning with the earliest history of freezing, the technology has been highly affected over the years by the developments and improvements in freezing techniques. In order to understand and handle the concepts associated with freezing of foods, it is necessary to examine the fundamental factors governing the freezing process.

Freezing Technology

Freezing has long been used as a method of preservation, and history reveals it was mostly shaped by the technological developments in the process. A small quantity of ice produced without using a "natural cold" in 1755 was regarded as the first milestone in the freezing process. Firstly, ice-salt systems were used to preserve fish and later on, by the late 1800's, freezing was introduced into large-scale operations as a method of commercial preservation. Meat, fish, and butter, the main products preserved in this early example, were frozen in storage chambers and handled as bulk commodities.

In the following years, scientists and researchers continuously worked to achieve success with commercial freezing trials on several food commodities. Among these commodities, fruits were one of the most important since freezing during the peak growing season had the advantage of preserving fruit for later processing into jams, jellies, ice cream, pies, and other bakery foods. Although commercial freezing of small fruits and berries first began around 1905 in the eastern part of the United States, the commercial freezing of vegetables is much more recent. Starting from 1917, only private firms conducted trials on freezing vegetables, but achieving good quality in frozen vegetables was not possible without pre-treatments due to the enzymatic deterioration. In 1929, the necessity of blanching to inactivate enzymes before freezing was concluded by several researchers to avoid deterioration and off-flavours caused by enzymatic degradation.

The modern freezing industry began in 1928 with the development of double-belt contact freezers by a technologist named Clarence Birdseye. After the revolution in the quick freezing process and equipment, the industry became more flexible, especially with the usage of multi-plate freezers. The earlier methods achieved successful freezing of fish and poultry, however with the new quick freezing system, packaged foods could be frozen between two metal belts as they moved through a freezing tunnel. This improvement was a great advantage in the commercial large-scale freezing of fruits and vegetables. Furthermore, quick-freezing of consumer-size packages helped frozen vegetables to be accepted rapidly in late 1930s.

Today, freezing is the only large-scale method that bridges the seasons, as well as variations in supply and demand of raw materials such as meat, fish, butter, fruits, and vegetables. Besides, it makes possible movement of large quantities of food over geographical distances. It is important to control the freezing process, including the pre-freezing preparation and post-freezing storage of the product, in order to achieve high-quality products. Therefore, the theory of the freezing process and the parameters involved should be understood clearly.

Freezing Process

The freezing process mainly consists of thermodynamic and kinetic factors, which can dominate each other at a particular stage in the freezing process. Major thermal events are accompanied by reduction in heat content of the material during the freezing process as is shown in figure. The material to be frozen first cools down to the temperature at which nucleation starts. Before ice can form, a nucleus, or a seed, is required upon which the crystal can grow; the process of producing this seed is defined as nucleation. Once the first crystal appears in the solution, a phase change occurs from liquid to solid with further crystal growth. Therefore, nucleation serves as the initial process of freezing, and can be considered as the critical step that results in a complete phase change.

Freezing Point of Foods

Freezing point is defined as the temperature at which the first ice crystal appears and the liquid at that temperature is in equilibrium with the solid. If the freezing point of pure water is considered, this temperature will correspond to 0°C (273°K). However, when food systems are frozen, the process becomes more complex due to the existence of both free and bound water. Bound water does not freeze even at very low temperatures. Unfreezable water contains soluble solids, which cause a decrease in the freezing point of water lower than 0°C. During the freezing process, the concentration of soluble solids increases in the unfrozen water, resulting in a variation in freezing temperature. Therefore, the temperature at which the first ice crystal appears is commonly regarded as the initial freezing temperature. There are empirical equations in literature that can calculate the initial freezing temperature of certain foods as a function of their moisture content.

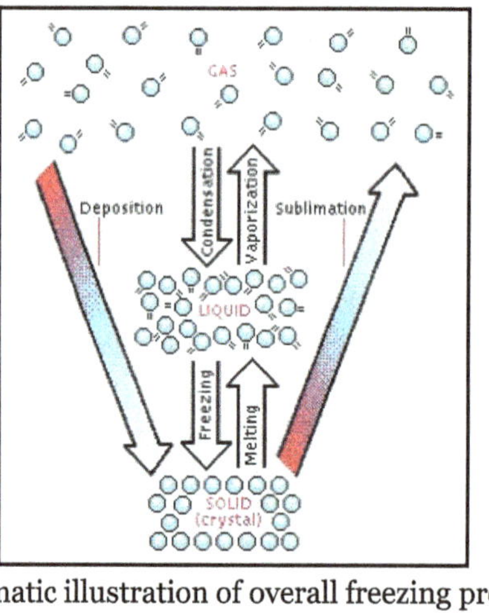

A schematic illustration of overall freezing process.

There are several methods of food freezing, and depending on the method used, the quality of the frozen food may vary. However, regardless of the method chosen, the main principle behind all freezing processes is the same in terms of process parameters. The International Institute of Refrigeration (IIR) has provided definitions to establish a basis for the freezing process. According to their definition, the freezing process is basically divided into three stages based on major temperature changes in a particular location in the product, as shown in figures for pure water and food respectively.

Beginning with the prefreezing stage, the food is subjected to the freezing process until the appearance of the first crystal. If the material frozen is pure water, the freezing temperature will be 0 °C and, up to this temperature, there will be a sub cooling until the ice formation begins. In the case of foods during this stage, the temperature decreases to below freezing temperature and, with the formation of the first ice crystal, increases to freezing temperature. The second stage is the freezing period; a phase change occurs, transforming water into ice. For pure water, temperature at this stage is constant; however, it decreases slightly in foods, due to the increasing concentration of solutes in the unfrozen water portion. The last stage starts when the product temperature reaches the point where most freezable water has been converted to ice, and ends when the temperature is reduced to storage temperature.

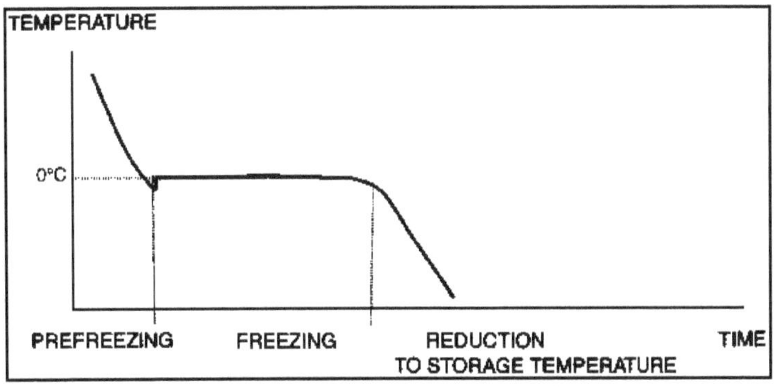

Practical definition of the freezing process for pure water.

The freezing time and freezing rate are the most important parameters in designing freezing systems. The quality of the frozen product is mostly affected by the rate of freezing, while time of freezing is calculated according to the rate of freezing. For industrial applications, they are the most essential parameters in the process when comparing different types of freezing systems and equipment.

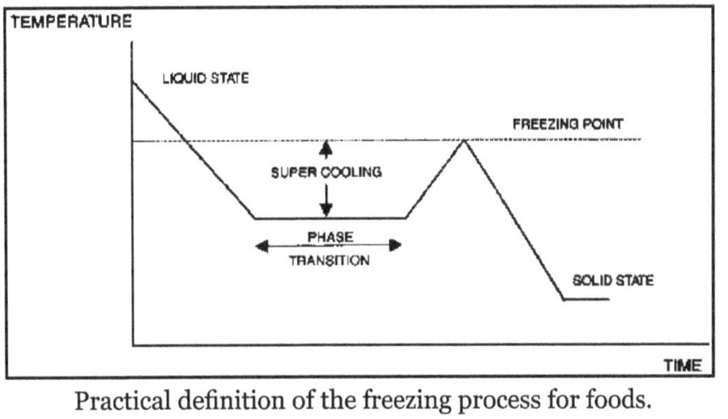

Practical definition of the freezing process for foods.

Freezing Time

Again, freezing time is one of the most important parameters in the freezing process, defined as time required to lower product temperature from its initial temperature to a given temperature at its thermal center. Since the temperature distribution within the product varies during freezing process, the thermal center is generally taken as reference. Thus, when the geometrical center of the product reaches the given final temperature, this ensures the average product temperature has been reduced to a storage value. Freezing time depends on several factors, including the initial and final temperatures of the product and the quantity of heat removed, as well as dimensions (especially thickness) and shape of product, heat transfer process, and temperature. The International Institute of Refrigeration (1986) defines various factors of freezing time in relation to both the product frozen and freezing equipment. The most important are:

- Dimensions and shape of product, particularly thickness;
- Initial and final temperatures;

- Temperature of refrigerating medium;
- Surface heat transfer coefficient of product;
- Change in enthalpy;
- Thermal conductivity of product.

Calculation of freezing time in food systems is difficult in comparison to pure systems since the freezing temperature changes continuously during the process. Using a simplified approach, time elapsed between initial freezing until when the entire product is frozen can be regarded as the freezing time. Plank's equation is commonly used to estimate freezing time, however due to assumptions involved in the calculation it is only useful for obtaining an approximation of freezing time. The derivation of the equation starts with the assumption the product being frozen is initially at freezing temperature. Therefore, the calculated freezing time represents only the freezing period. The equation can be further modified for different geometries including slab, cylinder, and sphere, where for each geometry, the coefficients are arranged in relation to the dimensions.

Geometry	P	R	Dimension
Infinite slab	1/2	1/8	thickness e
Infinite cylinder	1/4	1/16	radius r
Sphere	1/6	1/24	radius r

Coefficients P and R of Equation 1

$$t_F = \frac{\rho H}{T_F - T_e}\left[\frac{e^2 R}{k} + \frac{eP}{h}\right]$$

where l_1 is the latent heat of frozen fraction, k and r are the thermal conductivity and density of the frozen layer, while h is the coefficient of heat transfer by convection to the exterior. T_f denotes the body temperature of the product when introduced into a freezer in wich the external temperature is T_e The coefficients R and P are given in table and arranged according to the geometry of the product frozen. where the letter e denotes the dimension (i.e. for infinite slab geometry, e is thickness of the slab and for infinite cylinder or sphere e is replaced by r which denotes the radius of the clylinder or sphere).

As mentioned earlier, the equation of Plank assumes the food is at a freezing temperature at the beginning of the freezing process. However, the food is usually at a temperature higher than freezing temperature. The real freezing time should therefore be the sum of time calculated from the equation of Plank and the time needed for the product's surface to decrease from initial temperature to freezing temperature.

Several works have attempted to calculate real freezing time, as in one presented by Nagaoka *et al.* Nagaoka's equation calculates the amount of heat elimination required to decrease a product's temperature from initial temperature to freezing temperature, as well as the amount of heat released during the phase change and the amount of heat eliminated to reach freezing temperature. Further empirical equations can be found in literature in detail.

$$t_F = \frac{\rho \Delta H}{T_F - T_e}\left[\frac{Re^2}{k} + \frac{PI}{h}\right]\left[1 + 0.008\left(T_i - T_F\right)\right]$$

Where, T_i is the temperature of the food at the initiation of freezing, DH is the difference between the enthalpy of the food at initial temperature and end of freezing. Re and Pl are the dimensionless numbers, while k and h are the thermal conductivity and the coefficient of heat transfer, respectively.

For calculating freezing time of products with irregular shape, a common property of most food products - especially fruits and vegetables - a dimensionless factor has been employed in equations.

Freezing Rate

The freezing rate (°C/h) for a product or package is defined as the ratio of difference between initial and final temperature of product to freezing time. At a particular location within the product, a local freezing rate can be defined as the ratio of the difference between the initial temperature and desired temperature to the time elapsed in reaching the given final temperature. The quality of frozen products is largely dependent on the rate of freezing (Ramaswamy and Tung, 1984). Generally, rapid freezing results in better quality frozen products when compared with slow freezing. If freezing is instantaneous, there will be more locations within the food where crystallization begins. In contrast, if freezing is slow, the crystal growth will be slower with few nucleation sites resulting in larger ice crystals. Large ice crystals are known to cause mechanical damage to cell walls in addition to cell dehydration. Thus, the rate of freezing for plant tissues is extremely important due to the effect of freezing rate on the size of ice crystals, cell hydration, and damage to cell walls. The figure shows a general behavior of the dynamics curve of freezing preservation.

Rapid freezing is advantageous for freezing of many foods, however some products are susceptible to cracking when exposed to extremely low temperature for long periods. Several mechanisms, including volume expansion, contraction and expansion, and building of internal pressure, are proposed in literature explaining the mechanisms of product damage during freezing.

Energy Requirements

For fruits and vegetables, the amount of energy required for freezing is calculated based on the enthalpy change and the amount of product to be frozen. The following equation is reported by Riedel (1949) for calculation of refrigeration requirements for fruits and vegetables.

$$\Delta H = \left[1 - \frac{X_{SNJ}}{100}\right]\Delta H_j + 1.21\left[\frac{X_{SNJ}}{100}\right]\Delta T$$

X_{SNJ} : Percentage of the product solids different from juice (Dry matter fraction of the juice);

DH_j : Enthalpy change during freezing of the juice fraction;

DT : Temperature difference between initial and final temperature of the product.

Refrigeration

Refrigeration is defined as the elimination of heat from a material at a temperature higher than the temperature of its surroundings. The mechanism of refrigeration is a part of the freezing process and freezing storage involved in the thermodynamic aspects of freezing. According to the second law of thermodynamics, heat only flows from higher to lower temperatures. Therefore, in order to raise the heat from a lower to a higher temperature level, expenditure of work is needed. The aim of industrial refrigeration processes is to eliminate heat from low temperature points towards points with higher temperature. For this reason, either closed mechanical refrigeration cycles in which refrigeration fluids circulate, or open cryogenic systems with liquid nitrogen (LIN) or carbon dioxide (CO_2), are commonly used by the food industry.

The main elements in a closed mechanical refrigeration system are the condenser, compressor, evaporator, and the expansion valve. The refrigerants hydro chlorofluorocarbon (HCFC) and ammonia are examples of the refrigerants circulated in these types of mechanical refrigeration systems. A simple scheme for the closed mechanical refrigeration system.

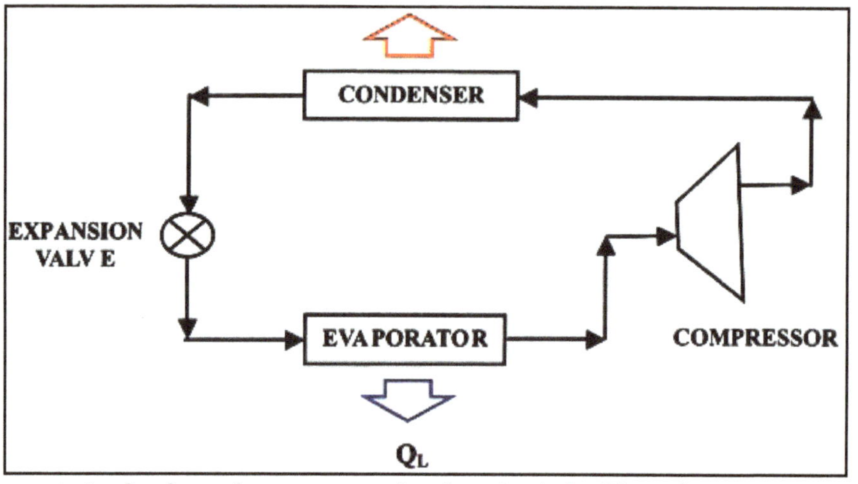

A simple scheme for a one-stage closed mechanical refrigeration system.

Starting at the suction point of the compressor, fluid in a vapor state is compressed into the compressor where an increase in temperature and pressure takes place. The fluid then flows through the condenser where it decreases in energy by giving off heat and converting to a liquid state. After the phase, a change occurs inside the condenser, the fluid flows through the expansion valve where the pressure decreases to convert liquid into a form of liquid-gas mixture. Finally, the liquid-gas mixture flows through the evaporator where it is converted into a saturated vapor state and removes heat from the environment in the process of cooling. With this last stage the loop restarts again.

The other refrigeration system employed by the food industry is the cryogenic system with carbon dioxide or liquid nitrogen. The refrigerant in this system is consumed differently from the circulating fluid in closed mechanical systems.

Refrigerants

There are several refrigerants available for refrigeration systems. The selection of a proper refrigerant is based on physical, thermodynamic, and chemical properties of the fluid. Environmental considerations are also important in refrigerant selection, since leaks within the system produce deleterious effects on the atmospheric ozone layer. Some refrigerants, including halocarbons, have been banned to avoid potential hazardous effects. For industrial applications, ammonia is commonly used, while chlorofluoromethane and tetrafluoroethane are also recommended as refrigerants.

Freezing Capacity

Freezing equipment selection is based on the requirements for freezing a certain quantity of food per hour. For any type of freezer, freezing capacity (expressed in tonnes per hour) is defined as the ratio of the quantity of the product that can be loaded into the freezer to the holding time of the product in that particular freezer. The first parameter, the amount of food product loaded into the freezer, is affected by both the dimensions of the product and the mechanical constraints of the freezer. The denominator (holding time) has an important role in freezing systems and is based on the calculation of the amount of heat removed from the product per hour, which varies depending on the type of product frozen.

Freezing Systems

There is a variety of freezing systems available for freezing, and for most products, more than one type of freezer can be used. Therefore, in selecting a freezing system initially, a cost-benefit analysis should be conducted based on three important factors: economics, functionality, and feasibility. Financial considerations mainly involve capital investment and the production cost of selected equipment. Product losses during freezing operation should be included in cost estimation since generating higher cost freezers may have other benefits in terms of reducing product losses. Functional factors are mostly based on the suitability of the selected freezer for particular products. The mode of process, either in-line or batch, should be considered based on the fact that computerized systems are becoming more important for ease of handling and lowering production costs. Mechanical constraints for the freezer should also be considered since some types of freezers are not physically suitable for freezing certain products. Lastly, the feasibility of the process should be considered in terms of plant location or location of the processing area, as well as cleanability and hygienic design, and desired product quality.

These factors and initial considerations can help eliminate several choices in freezer selection, but the relative importance of factors may change depending on the process. For developing countries where the freezing application is relatively new, the cost factor becomes more important than other factors due to the decreased production rates and need for lower capital investment costs.

Freezing Equipment

The industrial equipment for freezing can be categorized in many ways, namely as equipment used for batch or in-line operation, heat transfer systems (air, contact, cryogenic), and product stability. The rate of heat transfer from the freezing medium to the product is important in defining the freezing time of the product. Therefore, the equipment selected for freezing process characterizes the rate of freezing.

Air-blast Freezers

The air blast freezer is one the oldest and commonly used freezing equipment due to its temperature stability and versatility for several product types. In general, air is used as the freezing medium in the freezing design, either as still air or forced air. Freezing is accomplished by placing the food in freezing rooms called sharp freezers. Still, air freezing is the cheapest way of freezing and has the added advantage of a constant temperature during frozen storage, which allows usage for unprocessed bulk products like beef quarters and fish. However, it is the slowest method of freezing due to the low surface heat transfer coefficient of circulating air inside the room. Freezing time in sharp freezers is largely dependent on the temperature of the freezing chamber and the type, initial temperature, and size of product. An improved version of the still air freezer is the forced air freezer, which consists of air circulation by convection inside the freezing room. However, even modification of the sharp freezer with extra refrigeration capacity and fans for increased air circulation does not help control the air flow over the products during slow freezing. A typical design for air blast freezers.

There are a considerable number of designs and arrangements for air blast freezers, primarily grouped in two categories depending on the mode of process, as either inline or batch. Continuous freezers are the most suitable systems for mass production of packaged products with similar freezing times, in which the product is carried through on trucks or on conveyors. The system works on a semi-batch principle when trucks are used, since they remain stationary during the process except when a new truck enters one end of the tunnel, thus moving the others along to release a finished one at the exit. The batch freezers are more flexible since a variety of products can be frozen at the same time on individual trolleys. Over-loading may be a problem for these types of freezers, thus the process requires closer supervision than continuous systems.

Air blast freezer.

Tunnel Freezers

In tunnel freezers, the products on trays are placed in racks or trolleys and frozen with cold air circulation inside the tunnel. In order to allow air circulation, optimum space is provided between layers of trolley, which can be moved continuously in and out of the freezer manually or by forklift trucks. This freezing system is suitable for all types of products, although there are some mechanical constraints including the requirement of high manpower for handling, cleaning, and transportation of trays. A trolley for a tunnel freezer.

Trolley in a tunnel freezer.

Belt Freezers

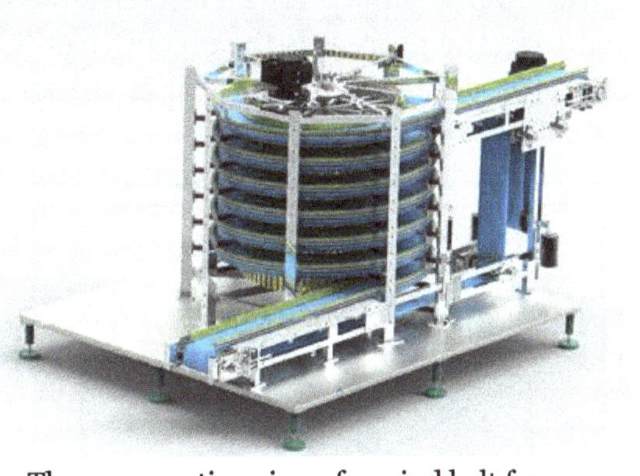

The cross-section view of a spiral belt freezer.

Belt freezers were first designed to provide continuous product flow with the help of a wire mesh conveyor inside the blast rooms. A poor heat transfer mechanism and the mechanical problems were solved in modern belt freezers by providing a vertical airflow to force air through the product layer. Airflow has good contact with the product only when the entire product is evenly distributed over the conveyor belt. In order to decrease required floor space, the belts can be arranged in a multi-tier belt freezer or a spiral belt freezer. Spiral belt freezers consist of a belt that can be bent laterally around a rotating drum to maximize belt surface area in a given floor space. This type of design has the advantage of eliminating product damage in transfer points, especially for products that require gentle handling. Both packed and unpacked products with long freezing times (10 min to 3 hr) can be frozen in spiral belt freezers due to the flexibility of the equipment. A typical spiral belt freezer.

Fluidized Bed Freezers

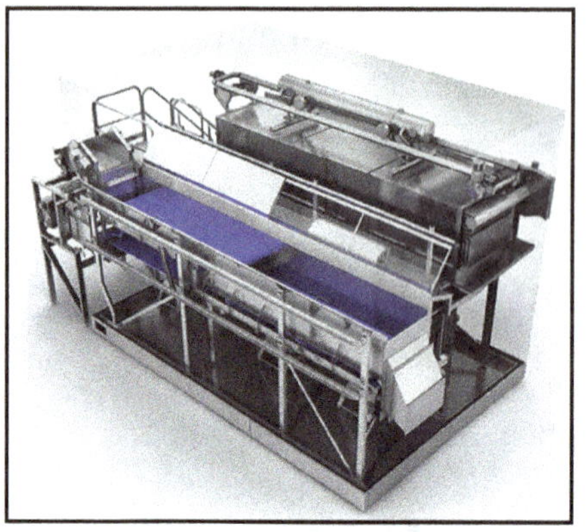

Cross-sectional view of a fluidized bed freezer.

The fluidized bed freezer, a fairly recent modified type of air-blast freezer for particular product types, consists of a bed with a perforated bottom through which cold air is blown vertically upwards. The system relies on forced cold air from beneath the conveyor belt, causing the products to suspend or float in the cold air stream. The use of high air velocity is very effective for freezing unpacked foods, especially when they can be completely surrounded by flowing air, as in the case of fluidized bed freezers.

The use of fluidization has several advantages compared with other methods of freezing since the product is individually quick frozen (IQF), which is convenient for particles with a tendency to stick together. The idea of individually quick frozen foods (IQF) started with the first technological developments aimed at quick freezing. The need for an effective means of freezing small particles with the potential for lumping during the process is the objective of IQF freezing. Small vegetables, prawns, shrimp, french-fried potatoes, diced meat, and fruits are some of the products now frozen with this technology.

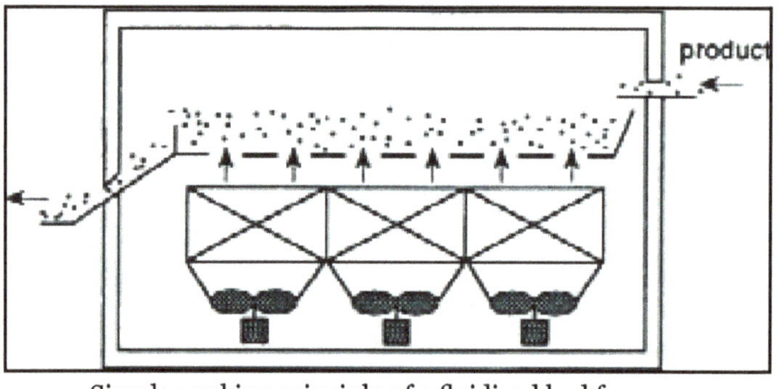

Simple working principle of a fluidized bed freezer.

Contact Freezers

Contact freezing is the one of the most efficient ways of freezing in terms of heat transfer mechanism. In the process of freezing, the product can be in direct or indirect contact with the freezing medium. For direct contact freezers, the product being frozen is fully surrounded by the freezing medium, the refrigerant, maximizing the heat transfer efficiency. A schematic illustration is given in figure. For indirect contact freezers, the product is indirectly exposed to the freezing medium while in contact with the belt or plate, which is in contact with the freezing medium.

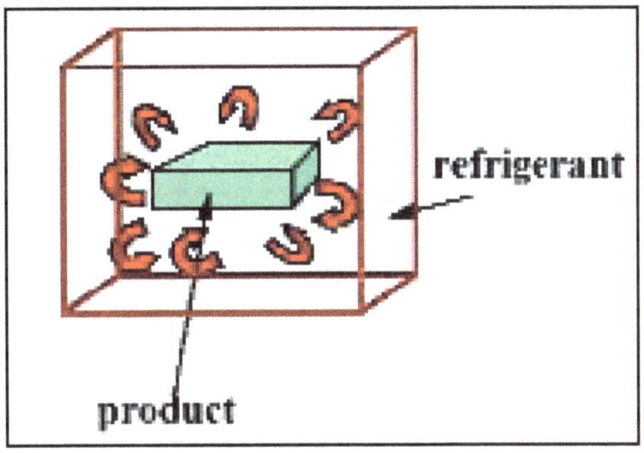

Direct contact freezer.

Immersion Freezers

The immersion freezer consists of a tank with a cooled freezing media, such as glycol, glycerol, sodium chloride, calcium chloride, and mixtures of salt and sugar. The product is immersed in this solution or sprayed while being conveyed through the freezer, resulting in fast temperature reduction through direct heat exchange. Direct immersion of a product into a liquid refrigerant is the most rapid way of freezing since liquids have better heat conducting properties than air. The solute used in the freezing system should be safe without taste, odour, colour, or flavour, and for successful freezing, products should be greater in density than the solution. Immersion freezing systems have been commonly used for shell freezing of large particles due to the reducing ability of product dehydration when the outer layer is frozen quickly. A commonly seen problem in these freezing systems is the dilution of solution with the product, which can change the concentration and process parameters. Thus, in order to avoid product contact with the liquid refrigerant, flexible membranes can be used.

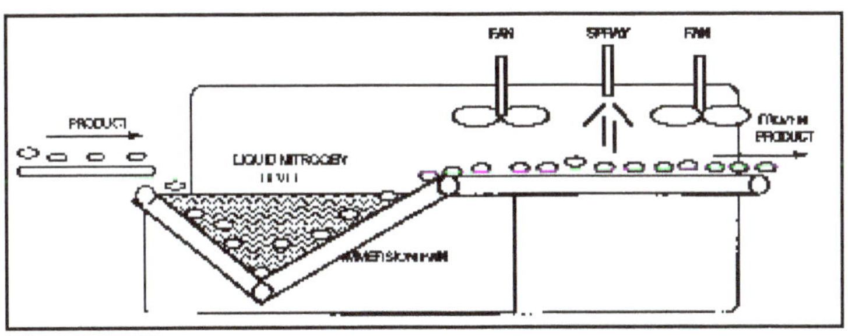

Simple illustration of a typical immersion freezer.

Indirect Contact Freezers

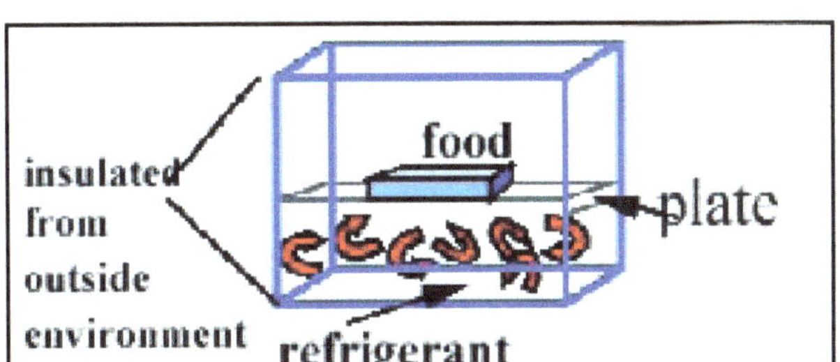

Indirect contact freezer.

In this type of freezer, materials being frozen are separated from the refrigerant by a conducting material, usually a steel plate. The mechanism of indirect contact freezer. Indirect contact freezers generally provide an efficient medium for heat transfer, although the system has some limitations, especially when used for packaged foods due to resistance of package to heat transfer. Additionally, corrosive effects may occur due to interaction of metal packages with heat transfer surfaces.

Plate Freezers

The most common type of contact freezer is the plate freezer. In this case, the product is pressed between hallow metal plates, either horizontally or vertically, with a refrigerant circulating inside the plates. Pressure is applied for good contact as schematically shown in figure.

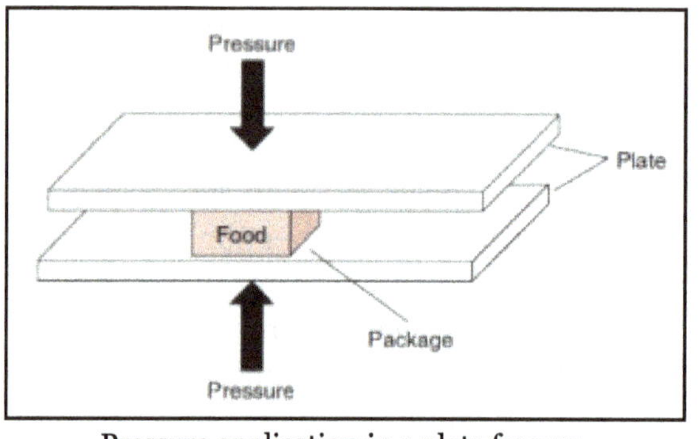

Pressure application in a plate freezer.

Plate freezer with a two-stage compressor and sea water condenser.

This type of freezing system is only limited to regular-shaped materials or blocks like beef patties or block-shaped packaged products.

Contact Belt Freezers

This type of freezer is designed with single-band or double-band for freezing of thin product layers as shown in figure. The design can be either straight forward or drum. Typical products frozen in belt freezers are, fruit pulps, egg yolk, sauces and soups.

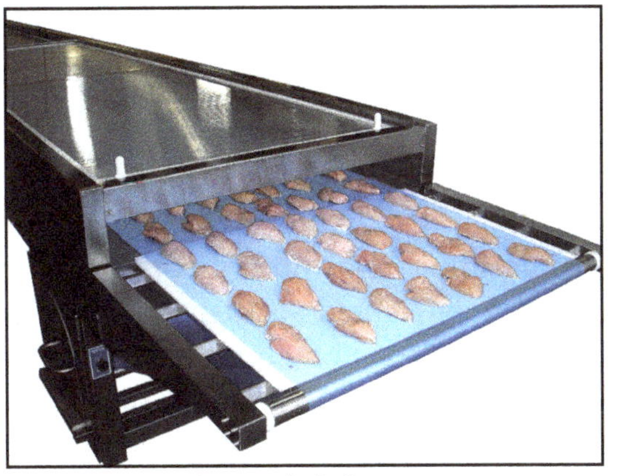

Contact belt freezer.

Cryogenic Freezers

Cryogenic freezing is a relatively new method of freezing in which the food is exposed to an atmosphere below -60°C through direct contact with liquefied gases such as nitrogen or carbon dioxide. This type of system differs from other freezing systems since it is not connected to a refrigeration plant; the refrigerants used are liquefied in large industrial installations and shipped to the food-freezing factory in pressure vessels. Thus, the small size and mobility of cryogenic freezers allow for flexibility in design and efficiency of the freezing application. Low initial investment and rather high operating costs are typical for cryogenic freezers.

Liquid Nitrogen Freezers

Liquid nitrogen, with a boiling temperature of -196°C at atmospheric pressure, is a by-product of oxygen manufacture. The refrigerant is sprayed into the freezer and evaporates both on leaving the spray nozzles and on contact with the products. The system is designed in a way that the refrigerant passes in counter current to the movement of the products on the belt giving high transfer efficiency. The refrigerant consumption is in the range of 1.2-kg refrigerant per kg of the product. Typical food products used in this system are, fish fillets, seafood, fruits, berries.

Liquid Carbon Dioxide Freezers

Liquid carbon dioxide exists as either a solid or gas when stored at atmospheric pressure. When the gas is released to the atmosphere at -70°C, half of the gas becomes dry-ice snow and the other half stays in the form of vapor. This unusual property of liquid carbon dioxide is used in a variety of freezing systems, one of which is a pre-freezing treatment before the product is exposed to nitrogen spray.

Packaging

Proper packaging of frozen food is important to protect the product from contamination and damage while in transit from the manufacturer to the consumer, as well as to preserve food value, flavour, colour, and texture. There are several factors considered in designing a suitable package for a frozen food. The package should be attractive to the consumer, protected from external contamination, and effective in terms of processing, handling, and cost. Proper selection is based on the type of package and material. There are typically three types of packaging used for frozen foods: primary, secondary, and tertiary. The primary package is in direct contact with the food and the food is kept inside the package up to the time of use. Secondary packaging is a form of multiple packaging used to handle packages together for sale. Tertiary packaging is used for bulk transportation of products.

Packaging materials should be moisture-vapor-proof to prevent evaporation, thus retaining the highest quality in frozen foods. Oxygen should also be completely evacuated from the package using a vacuum or gas-flush system to prevent migration of moisture and oxygen. Glass and rigid plastic are examples of moisture-vapor-proof packaging materials. Many packaging materials, however, are not moisture-vapor-proof, but are sufficiently moisture-vapor-resistant to retain satisfactory quality in foods. Most bags, wrapping materials, and waxed cartons used in freezing packaging are moisture-vapor-resistant. In general, the containers should be leakage free while easy to seal. Durability of the material is another important factor to consider, since the packaging material must not become brittle at low temperatures and crack.

A range of different packaging materials, mainly grouped as rigid and non-rigid containers, can be used for primary packaging. Glass, plastic, tin, and heavily waxed cardboard materials are in the rigid container group and usually used for packaging of liquid food products. Glass containers are mostly used for fruits and vegetables if they are not water-packed. Plastics are the derivatives of the oil-cracking industry. Non-rigid containers include bags and sheets made of moisture-vapor-resistant heavy aluminum foil, polyethylene or laminated papers. Bags are the most commonly used packaging materials for frozen fruits and vegetables due to their flexibility during processing and handling. They can be used with or without outer cardboard cartons to protect against tearing.

Shape and size of the container are also important factors in freezing products. Serving size may vary depending on the type of product and selection should be based on the amount of food determined for one meal. For shape of the container, freezer space must be considered since rigid containers with flat tops and bottoms stack well in the freezer, while round containers waste freezer space.

Frozen Storage and Distribution

The quality of the final product depends on the history of the raw material. Using the lowest possible temperature is essential for frozen storage, transport, and distribution in achieving a high-quality product, since deteriorative processes are mainly temperature dependent. The lower the product temperature is, the slower the speed of reaction is leading to loss of quality. The temperatures of supply chains in freezing applications from the factory to the retail cabinet should be carefully monitored. The temperature regime covering the freezing process, the cold-store temperatures (£ -18°C), distribution temperatures (£ -15°C), and retail display (£ -12°C) are given as legal standards.

Freezing Fruits and Vegetables in Small and Medium Scale Operations and its Potential Applications in Warm Climates

The preservation of fruits and vegetables by freezing is one of the most important methods for retaining high quality in agricultural products over long-term storage. In particular, the freshness qualities of raw fruits and vegetables can be retained for long periods, extending well beyond the normal season of most horticultural crops. The potential application of freezing preservation of fruits and vegetables, including tropical products, has been increasing recently in parallel with developments in developing countries. Freezing of fruits and vegetables in small and medium scale operations is detailed in the following topics and a general flowchart.

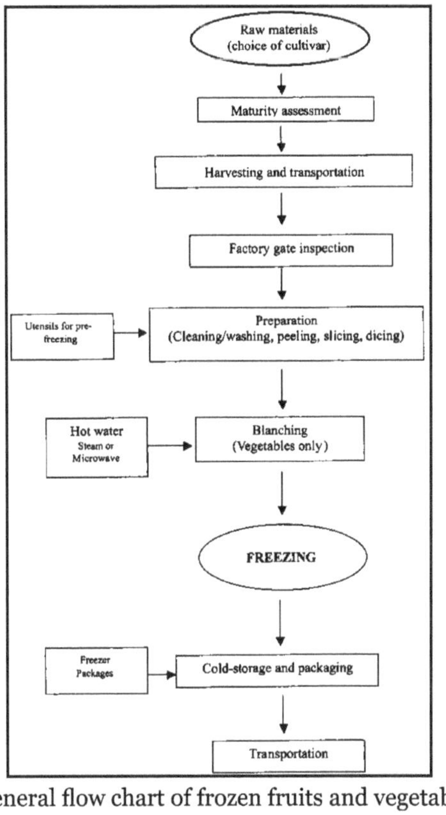

A general flow chart of frozen fruits and vegetables.

Freezing Fruits

The effect of freezing, frozen storage, and thawing on fruit quality has been investigated over several decades. Today frozen fruits constitute a large and important food group. The quality demanded in frozen fruit products is mostly based on the intended use of the product. If the fruit is to be eaten without any further processing after thawing, texture characteristics are more important when compared to use as a raw material in other industries. In general, conventional methods of freezing tend to destroy the turgidity of living cells in fruit tissue. Different from vegetables, fruits do not have a fibrous structure that can resist this destructive effect. Additionally, fruits to be frozen are harvested in a fully ripe state and are soft in texture. On the contrary, a great number of vegetables are frozen in an immature state. Fruits have delicate flavours that are easily damaged or changed by heat, indicating they are best eaten when raw and decrease in quality with processing. In the same way, attractive colour is important for frozen fruits. Chemical treatments or additives are often used to inactivate the deteriorative enzymes in fruits. Therefore, proper processing is essential for all steps involved, from harvesting to packaging and distribution.

Production and Harvesting

The characteristics of raw materials are of primary importance in determining the quality of the frozen product. These characteristics include several factors such as genetic makeup, climate of the growing area, type of fertilization, and maturity of harvest.

The ability to withstand rough handling, resistance to virus diseases, molds, uniformity in ripening, and yield are some of the important characteristics of fruits in terms of economical aspects considered in production. The use of mechanical harvesting generally causes bruising of fruits and results in a wide range of maturity levels for fruits. In contrast, hand-picking provides gentler handling and maturity sorting of fruits. However in most cases, it is non-economical compared to mechanical harvesting due to high labor cost.

As a rule, harvesting of fruits at an optimum level for commercial use is difficult. Simple tests like pressure tests are applied to determine when a fruit has reached optimum maturity for harvest. Colour is also one of the characteristics used in determining maturity since increased maturation causes a darker colour in fruits. A combination of colour and pressure tests is a better way to assess maturity level for harvesting.

Controlled atmosphere storage is a common method of storage for some fruits prior to freezing. In principle, a controlled atmosphere high in carbon dioxide and low in oxygen content slows down the rate of respiration, which may extend shelf life of any respiring fruit during storage. Due to the fact that these fruits do not ripen appreciably after picking, most fruits are picked as near to eating-ripe maturity as possible.

Pre-process Handling and Operations

Freezing preservation of fruits can only help retain the inherent quality present initially in a product since the process does not improve the quality characteristics of raw materials. Therefore, quality level of the raw materials prior to freezing is the major consideration for successful freezing. Washing and cutting generally results in losses when applied after thawing. Thus, fruits should be prepared prior to the freezing process in terms of peeling, slicing or cutting. Freezing preservation does not require specific unit operations for cleaning, rinsing, sorting, peeling, and cutting of fruits.

Fruits that require peeling before consumption should be peeled prior to freezing. Peeling is done by scalding the fruit in hot water, steam or hot lye solutions (Boyle and Wolford, 1968). The effect of peeling on the quality of frozen products has been studied for several fruits, including kiwi, banana, and mang. The rate of freezing can be increased by decreasing the size of products frozen, especially for large fruits. An increase in the freezing rate results in smaller ice crystals, which decreases cellular damage in fruit tissue. Banana, tomato, mango, and kiwi are some examples of large fruits commonly cut into smaller cubes or slices prior to freezing.

The objective of blanching is to inactivate the enzymes causing detrimental changes in colour, odour, flavour, and nutritive value, but heat treatment causes loss of such characteristics in fruits. Therefore, only a few types of fruits are blanched for inactivation of enzymes prior to freezing. The loss of water-soluble minerals and vitamins during blanching should also be minimized by keeping blanching time and temperature at an optimum combination.

Effect of Ingredients

Addition of sugars is an extremely important pretreatment for fruits prior to freezing since the treatment has the effect of excluding oxygen from the fruit, which helps to retain colour and appearance. Sugars when dissolved in solutions act by withdrawing water from cells by osmosis, resulting in very concentrated solutions inside the cells. The high concentration of solutes depresses the freezing point and therefore reduces the freezing within the cells, which inhibits excessive structural damage. Sugar syrups in the range of 30-60 percent sugar content are commonly used to cover the fruit completely, acting as a barrier to oxygen transmission and browning. Several experiments have shown the protective effect of sugar on flavour, odour, colour, and nutritive value during freezing, especially for frozen berries.

Packaging

Fruits exposed to oxygen are susceptible to oxidative degradation, resulting in browning and reduced storage life of products. Therefore, packaging of frozen fruits is based on excluding air from the fruit tissue. Replacement of oxygen with sugar solution or inert gas, consuming the oxygen by glucose-oxidase and/or the use of vacuum and oxygen-impermeable films are some of the methods currently employed for packaging frozen fruits. Plastic bags, plastic pots, paper bags, and cans are some of the most commonly used packaging materials (with or without oxygen removal) selected, based on penetration properties and thickness.

There are several types of fruit packs suitable for freezing: syrup pack, sugar pack, unsweetened pack, and tray pack and sugar replacement pack. The type of pack is usually selected according to the intended use for the fruit. Syrup-packed fruits are generally used for cooking purposes, while dry-packed and tray-packed fruits are good for serving raw in salads and garnishes.

Syrup Pack

The proportion of sugar to water used in a syrup pack depends on the sweetness of the fruit and the taste preference of the consumer. For most fruits, 40 percent sugar syrup is recommended. Lighter syrups are lower in calories and mostly desirable for mild-flavoured fruits to prevent masking the flavour, while heavier syrups may be used for very sour fruits.

Syrup is prepared by dissolving the sugar in warm water and cooling the solution down before usage. Just enough cooled syrup is used to cover the prepared fruit after it has been settled by jarring the container. In order to keep the fruit under the syrup, a small piece of crumpled waxed paper or other water resistant wrapping material is placed on top; the fruit is pressed down into the syrup before closing, then sealed and frozen.

Pectin can be used to reduce sugar content in syrups when freezing berries, cherries, and peaches. Pectin syrups are prepared by dissolving 1 box of powdered pectin with 1 cup of water. The solution is stirred and boiled for 1 minute; 1/2 cup of sugar is added and dissolved; the solution is then cooled down with the addition of cold water. Previously prepared fruit is put into a 4 to 6 quart bowl and enough pectin syrup is added to cover the fruit with a thin film. The pack is sealed and promptly frozen.

Sugar Packs

In preparing a sugar pack, sugar is first sprinkled over the fruit. Then the container is agitated gently until the juice is drawn out and the sugar is dissolved. This type of pack is generally used for soft sliced fruits such as peaches, strawberries, plums, and cherries, by using sufficient syrup to cover the fruit. Some whole fruits may also be coated with sugar prior to freezing.

Unsweetened Packs

Unsweetened packs can be prepared in several ways, either dry-packed, covered with water containing ascorbic acid, or packed in unsweetened juice. When water or juice is used in syrup and sugar packs, fruit is submerged by using a small piece of crumpled water-resistant material. Generally, unsweetened packs yield a lower quality product when compared with sugar packs, with the exception, some fruits such as raspberries, blueberries, scalded apples, gooseberries, currants, and cranberries maintain good quality without sugar.

Tray Packs

Unsweetened packs are generally prepared by using tray packs in which a single layer of prepared fruit is spread on shallow trays, frozen, and packaged in freezer bags promptly. The fruit sections remain loose without clumping together, which offers the advantage of using frozen fruit piece by piece.

Sugar Replacement Packs

Artificial sweeteners can be used instead of sugar in the form of sugar substitutes. The sweet taste of sugar can be replaced by using these kinds of sweeteners, however the beneficial effects of sugar like colour protection and thick syrup can not be replaced. Fruits frozen with sugar substitutes will freeze harder and thaw more slowly than fruits preserved with sugar.

Freezing Vegetables

Freezing is often considered the simplest and most natural way of preservation for vegetables. Frozen vegetables and potatoes form a significant proportion of the market in terms of frozen food consumption. The quality of frozen vegetables depends on the quality of fresh products, since freezing does not improve product quality. Pre-process handling, from the time vegetables are picked until ready to eat, is one of the major concerns in quality retention.

Fruit	Preparation	Type of Pack
Apples	Wash, peel, and slice into antidarkening solution - 3 tablespoons lemon juice per quart of water	Pack in 30-40% syrup, adding 1/2 teaspoon crystalline ascorbic acid per quart of syrup. Pack dry or with up to 1/2 cup sugar per quart of apple slices.
Apricots	Wash, halve, and pit. Peel and slice if desired. If apricots are not peeled, heat in boiling water for 1/2 minute to keep skins from toughening during freezing. Cool in cold water, drain.	Pack in 40% syrup, adding 3/4 teaspoon crystalline ascorbic acid per quart of syrup.
Avocados	Peel soft, ripe avocados. Cut in half, remove pit, mash pulp.	Add 1/8 teaspoon crystalline ascorbic acid to each quart of puree. Package in recipe-size amounts.
Berries	Select firm, fully ripe berries. Sort, wash, and drain.	Use 30% syrup pack, dry unsweetened pack, dry sugar pack, (3/4 cup sugar per quart of berries), or tray pack.

Cherries(sour or sweet)	Select well-colored, tree-ripened cherries. Stem, sort, and wash thoroughly. Drain and pit.	Pack in 30-40% syrup. Add 1/2 teaspoon ascorbic acid per quart of syrup. For pies and other cooked products, pack in dry sugar using 3/4-cup sugar per quart of fruit.
Citrus fruits, (sections or slices)	Select firm fruit, free of soft spots. Wash and peel.	Pack in 40% syrup or in fruit juice. Add 1/2 teaspoon ascorbic acid per quart of syrup or juice.
Grapes	Select firm, ripe grapes. Wash and remove stems. Leave seedless grapes whole. Cut grapes with seeds in half and remove seeds.	Pack in 20% syrup or pack without sugar. Use dry pack for halved grapes and tray pack for whole grapes.
Melons (cantaloupe, watermelon)	Select firm-fleshed, well-colored, ripe melons. Wash rinds well. Slice or cut into chunks.	Pack in 30% syrup or pack dry using no sugar. Pulp also may be crushed (except watermelon), adding 1 tablespoon sugar per quart. Freeze in recipe-size containers.

Crop Cultivar, Production and Maturity

The choice of the right cultivar and maturity before crop is harvested are the two most important factors affecting raw material quality. Raw material characteristics are usually related to the vegetable cultivar, crop production, crop maturity, harvesting practices, crop storage, transport, and factory reception.

The choice of crop cultivars is mostly based on their suitability for frozen preservation in terms of factory yield and product quality. Some of the characteristics used as selection criteria are as follows:

- Suitability for mechanical harvesting.
- Uniform maturity.
- Exceptional flavour and uniform colour and desirable texture.
- Resistance to diseases.
- High yield.

Although cultivar selection is a major factor affecting the quality of the final product, many practices in the field and factors during growth of crop can also have a significant effect on quality. Those practices include site selection for growth, nutrition of crop, and use of agricultural chemicals to control pests or diseases. The maturity assessment for harvesting is one of the most difficult parts of the production. In addition to conventional methods, new instruments and tests have been developed to predict the maturity of crops that help determining the optimum harvest time, although the maturity assessment differs according to crop variety.

Harvesting

At optimum maturity, physiological changes in several vegetables take place very rapidly. Thus, the determination of optimum harvesting time is critical. Some vegetables such as green peas and sweet corn only have a short period during which they are of prime quality. If harvesting is delayed beyond this point, quality deteriorates and the crop may quickly become unacceptable. Most of the vegetables are subjected to bruising during harvesting.

Pre-process Handling

Vegetables at peak flavour and texture are used for freezing. Postharvest delays in handling vegetables are known to produce deterioration in flavour, texture, colour, and nutrients. Therefore, the delays between harvest and processing should be reduced to retain fresh quality prior to freezing. Cooling vegetables by cold water, air blasting, or ice will often reduce the rate of post-harvest losses sufficiently, providing extra hours of high quality retention for transporting raw material to considerable distances from the field to the processing plant.

Blanching

Blanching is the exposure of the vegetables to boiling water or steam for a brief period of time to inactivate enzymes. Practically every vegetable (except herbs and green peppers) needs to be blanched and promptly cooled prior to freezing, since heating slows or stops the enzyme action, which causes vegetables to grow and mature. After maturation, however, enzymes can cause loss in quality, flavour, colour, texture, and nutrients. If vegetables are not heated sufficiently, the enzymes will continue to be active during frozen storage and may cause the vegetables to toughen or develop off-flavours and colours. Blanching also causes wilting or softening of vegetables, making them easier to pack. It destroys some bacteria and helps remove any surface dirt.

Blanching in hot water at 70 to 105°C has been associated with the destruction of enzyme activity. Blanching is usually carried out between 75 and 95°C for 1 to 10 minutes, depending on the size of individual vegetable pieces. Blanched vegetables should be promptly cooled down to control and minimize the degradation of soluble and heat-labile nutrients.

The enzymes used as indicators of effectiveness of the blanching treatment are peroxidase, catalase, and more recently lipoxygenase. Peroxidase inactivation is commonly used in vegetable processing, since peroxidase is easily detected and is the most heat stable of these enzymes.

Vegetables can be blanched in hot water, steam, and in the microwave. Hot water blanching is the most common way of processing vegetables. Blanching times recommended for various vegetables are given in table, which indicates that the operation time can vary depending on the intended product use. For water blanching, vegetables are put in a basket and then placed in a kettle of boiling water covered with a lid. Timing begins immediately. Steam blanching takes longer than the water method, but helps retain water-soluble nutrients such as water-soluble vitamins. For steam blanching, a single layer of vegetables is placed on a rack or in a basket at 3-5 cm above water boiling in a kettle. A tightly fitted lid is placed on the kettle and timing is started. Microwave blanching is usually recommended for small quantities of vegetables prior to freezing. Due to the non-uniform heating disadvantage of microwaves, research is still being conducted to obtain better results with microwave blanching.

Vegetable	Preparation	Blanch/Freeze	
Asparagus	Wash and sort by sizeSnap off tough ends. Cut stalks into 5-cm lengths.	Water blanch:	2 min
		Steam blanch:	3 min
Beans	Wash and trim the ends. Cut if desired.	Water blanch:	Steam blanch:
		Whole: 3 min.	Whole: 4 min.
		Cut: 2min.	Cut: 3min.

Beets	Wash and remove the tops leaving 2.5 cm of stem and root.	Cook until tender: 25-30 min Cool promptly, peel, trim. Cut into slices or cubes and pack.	
Broccoli	Wash and cut into pieces.	Water blanch:	3 min.
		Steam blanch:	3 min.
Cabbage	Wash and cut into wedges.	Water blanch:	3 min.
		Steam blanch:	4 min.
Carrots	Wash, peel and trim. Cut if desired.	Water blanch: 5 min.	
Cauliflower	Discard leaves; steam and wash. Break into flowerets.	Water blanch:	Steam blanch:
		Whole: 5 min.	Whole: 7 min
Corn	Remove husks and silks. Trim ends and wash.	Water blanch:	Steam blanch:
		Whole: 5 min.	Whole: 7 min
Greens	Select young tender greens. Wash and trim the leaves.	Water blanch:	2 min.
		Steam blanch:	3 min.
Herbs	Wash.	No heat treatment is needed.	
Mushrooms	Wipe and damp with paper towel. Trim hard tip of stems. Sort and cut large mushrooms.	May be frozen without heat treatment.	
Peas	Shell garden peas.	Water blanch:	Steam blanch:
		1-1/2 min.	1-1/2 min.
Peppers	Wash, remove stems and seeds.	Freeze whole or cut as desired. No heat treatment is needed.	
Potatoes	Peel, cut or grate as desired.	Water blanch:	
		Whole: 5 min.	
		Pieces: 2-3 min.	

Packaging

There are several factors to consider in packaging frozen vegetables, which include protection from atmospheric oxygen, prevention of moisture loss, retention of flavour, and rate of heat transfer through the package . There are two basic packing methods recommended for frozen vegetables: dry pack and tray pack.

In the dry pack method, the blanched and drained vegetables are put into meal-sized freezer bags and packed tightly to cut down on the amount of air in the package. Proper headspace (approximately 2 cm) is left at the top of rigid containers before closing. For freezer bags, the headspace is larger. Provision for headspace is not necessary for foods such as broccoli, asparagus, and brussels sprouts, as they do not pack tightly in containers.

In the tray pack method, chilled, well-drained vegetables are placed in a single layer on shallow trays or pans. Trays are placed in a freezer until the vegetables become firm, then removed. Vegetables are filled into containers. Tray-packed foods do not freeze in a block but remain loosely distributed so that the amount needed can be poured from the container and the package reclosed.

Physical Aspects of Freezing

Moisture migration is the principal physical change occurring in frozen foods, affecting the physical, chemical, and biochemical properties, including texture and palatability of the food.

Texture

Most fruits and vegetables are over 90 percent water of total weight. The water and dissolved solutes inside the rigid plant cell walls give support to the plant structure, and texture to the fruit or vegetable tissue. In the process of freezing, when water in the cells freezes, an expansion occurs and ice crystals cause the cell walls to rupture. Consequently, the texture of the produce is generally much softer after thawing when compared to non-frozen produce. This textural difference is especially noticeable in products normally consumed raw, as in the case of fruits. It is usually recommended that frozen fruits be served before they are completely thawed, since in the partially thawed state the effect of freezing on the fruit tissue is less noticeable. On the other hand, due to the fact cooking also softens cell walls, textural changes caused by freezing are not significantly noticeable in products cooked before eating, as in the case of most vegetables.

Freezer Burn

One of the most common forms of quality degradation due to moisture migration in frozen foods is freezer burn, a condition defined as the glassy appearance in some frozen products produced by ice crystals evaporating on the surface area of a product. The grainy, brownish spots occurring on the product cause the tissue to become dry and tough and to develop off-flavors. This quality defect can be prevented by using heavyweight, moisture proof packaging during the freezing process.

Chemical Aspects of Freezing

Chemical changes that can cause spoilage and deterioration of fresh fruits and vegetables will continue after harvesting. This is the main reason for eliminating any delays during pre-freezing operations. As mentioned earlier, several enzymes that cause the loss of color, loss of nutrients, flavor changes, and color changes in frozen fruits and vegetables, should be inactivated by means of thermal treatments prior to freezing. In most cases, blanching is essential for producing quality frozen vegetables, since it also helps destroy microorganisms on the surface of the produce. However, in processing fruits, heat treatment may cause more degradation in quality. In this case, enzymes in frozen fruits can be controlled by using chemical compounds, which interfere with deteriorative chemical reactions. Ascorbic acid is an example that may be used in its pure form or in commercial mixtures with sugars for inhibition of enzymes in fruits.

Development of rancid oxidative flavors through contact of the frozen product with air is another group of chemical changes that can take place in frozen products. This problem can be controlled by excluding oxygen through proper packaging as mentioned earlier. It is also advisable to remove as much air as possible from the freezer bag or container to reduce the amount of air in contact with the product.

Formulation and Processing of Selected Frozen Food Prototypes

Selecting a Formulation for Mixed Fruits

One of the most common commercially frozen fruit products is mixed frozen berries. The mixtures typically contain combinations of raspberries, blackberries, blueberries, and strawberries. There is a wide range of mixtures available in the market for frozen berry combinations. Raspberries and blackberries for example, which are known to freeze well and retain their wholeness and shape, dependent on the structure of the fruit, are strongly associated with their cultivar. The processing requirements for different varieties of berries do not change significantly. Therefore, a mixture of raspberries and blackberries is chosen in this case as a fruit formulation to simplify the freezing process.

Raspberries and blackberries.

Procedure for processing mixed berries

- Full-flavoured, ripe berries of like size preferably with tender skins are selected;
- Berries are sorted, washed, and drained;
- Berries are packed into containers and covered with cold 40 percent sugar syrup, with proper headspace;
- Polyethylene freezer bags are sealed and frozen.

Selecting a Formulation for Mixed Vegetables

Mixed vegetables

Frozen mixed vegetables constitute a large portion of the frozen vegetable market and are now available in an ever-increasing variety of mixtures. The mixtures include three or more types of vegetables, properly prepared and blanched. The USDA standards for frozen mixed vegetables describe this item as a mixture containing three or more of the basic vegetables - beans, carrots, corn, and peas. When three vegetables are used, none of the vegetables should be more than 40 percent of the total weight; the individual percent decreases with increased number of vegetable types.

In a mixed frozen vegetable product, vegetables of different sizes are present in the mixture. Therefore, during pre-freeze treatments, especially blanching, care must be taken to be sure all vegetables are blanched properly.

Procedure for processing mixed vegetables:

- A mixture of four vegetables in which none of the vegetables is less than 8 percent by weight nor more than 35 percent by weight of all the frozen mixed vegetables are selected;
- The vegetables are sorted, washed and peeled;
- They are cut into uniform size and blanched in hot water for 5 minutes, and immediately cooled after blanching;
- Packed and frozen.

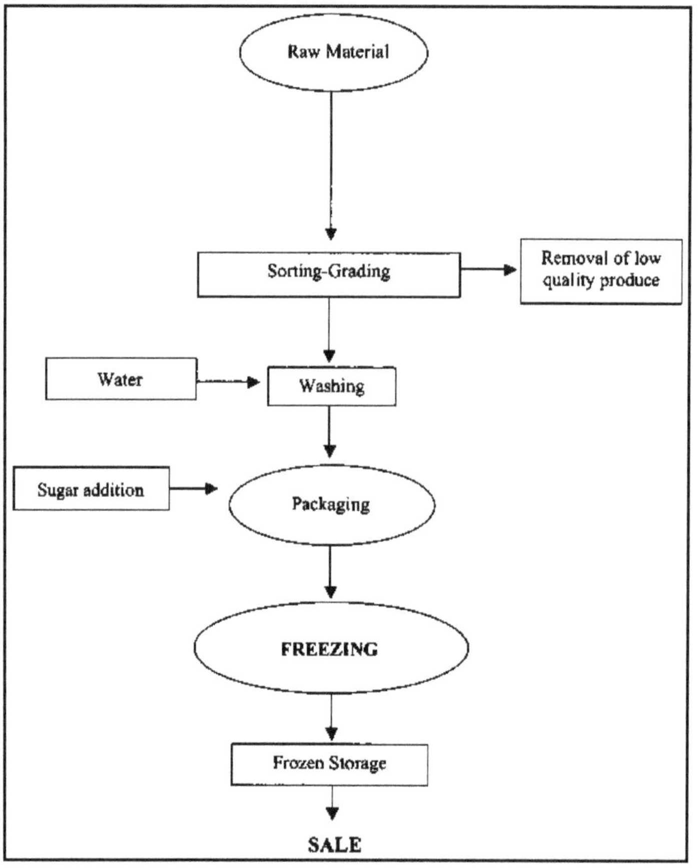

Flow diagram of freezing process for fruit-based product.

Boiling

Boiling is the process of applying heat to water until the temperature reaches about 100°C. Boiling foods in water cannot completely destroy all microorganisms, but the vegetative cells of bacteria, yeasts and moulds are generally quickly destroyed at temperatures of 100°C or above. Spores of some bacteria are extremely resistant to heat and are not killed at this temperature, although their growth is prevented. For this reason, boiling food can rarely be relied upon to ensure complete destruction of all organisms. However, most pathogens are killed, provided that sufficient exposure time is maintained. Although the spores of Clostridium botulinum, which causes botulism, are extremely heat-resistant, the toxin produced by this organism is readily destroyed by boiling. However, some toxins produced by other bacteria such as staphylococci are not easily inactivated. Thermophilic (heat-loving) organisms may survive the effects of boiling and can cause food spoilage if environmental conditions are favourable for them.

Types

Nucleate

Nucleate boiling is characterized by the growth of bubbles or pops on a heated surface, which rises from discrete points on a surface, whose temperature is only slightly above the liquids. In general, the number of nucleation sites are increased by an increasing surface temperature.

Nucleate boiling of water over a kitchen stove burner.

An irregular surface of the boiling vessel (i.e., increased surface roughness) or additives to the fluid (i.e., surfactants and/or nanoparticles) can create additional nucleation sites, while an exceptionally smooth surface, such as plastic, lends itself to superheating. Under these conditions, a heated liquid may show boiling delay and the temperature may go somewhat above the boiling point without boiling.

Critical Heat Flux

As the boiling surface is heated above a critical temperature, a film of vapor forms on the surface. Since this vapor film is much less capable of carrying heat away from the surface, the temperature rises very rapidly beyond this point into the transition boiling regime. The point at which this occurs is dependent on the characteristics of boiling fluid and the heating surface in question.

Transition

Transition boiling may be defined as the unstable boiling, which occurs at surface temperatures between the maximum attainable in nucleate and the minimum attainable in film boiling.

The formation of bubbles in a heated liquid is a complex physical process which often involves cavitation and acoustic effects, such as the broad-spectrum hiss one hears in a kettle not yet heated to the point where bubbles boil to the surface.

Film

If a surface heating the liquid is significantly hotter than the liquid then film boiling will occur, where a thin layer of vapor, which has low thermal conductivity, insulates the surface. This condition of a vapor film insulating the surface from the liquid characterizes *film boiling*.

Uses

In Cooking

Boiling is the method of cooking food in boiling water or other water-based liquids such as stock or milk. Simmering is gentle boiling, while in poaching the cooking liquid moves but scarcely bubbles.

Boiling pasta

The boiling point of water is typically considered to be 100°C or 212°F. Pressure and a change in the composition of the liquid may alter the boiling point of the liquid. For this reason, high elevation cooking generally takes longer since boiling point is a function of atmospheric pressure. In Denver, Colorado, USA, which is at an elevation of about one mile, water boils at approximately 95°C or 203°F. Depending on the type of food and the elevation, the boiling water may not be hot enough to cook the food properly. Similarly, increasing the pressure as in a pressure cooker raises the temperature of the contents above the open air boiling point.

Some science suggests adding a water-soluble substance, such as salt or sugar also increases the boiling point. This is called boiling-point elevation. At palatable concentrations of salt, the effect is very small, and the boiling point elevation is difficult to notice and this is why experiments to prove this are considered inconclusive. However, while making thick sugar syrup, such as for Gulab Jamun, one will notice boiling point elevation. Due to variations in composition and pressure, the boiling point of water is almost never exactly 100 °C, but rather close enough for cooking.

Boiling milk

Foods suitable for boiling include vegetables, starchy foods such as rice, noodles and potatoes, eggs, meats, sauces, stocks, and soups.

Boiling has several advantages. It is safe and simple, and it is appropriate for large-scale cookery. Older, tougher, cheaper cuts of meat and poultry can be made digestible. Nutritious, well-flavored stock is produced. Also, maximum color and nutritive value is retained when cooking green vegetables, provided boiling time is kept to the minimum.

Boiling food over a fireplace.

On the other hand, there are several disadvantages. There is a loss of soluble vitamins from foods to the water (if the water is discarded). Boiling can also be a slow method of cooking food.

Boiling can be done in several ways: The food can be placed into already rapidly boiling water and left to cook, the heat can be turned down and the food can be simmered or the food can also be placed into the pot, and cold water may be added to the pot. This may then be boiled until the food is satisfactory.

Water on the outside of a pot, i.e., a wet pot, increases the time it takes the pot of water to boil. The pot will heat at a normal rate once all excess water on the outside of the pot evaporates.

Boiling is also often used to remove salt from certain foodstuffs, such as bacon if a less saline product is required.

Boil-in-the-bag

Also known as "boil-in-bag", this involves heating or cooking ready-made foods sealed in a thick plastic bag. The bag containing the food, often frozen, is submerged in boiling water for a prescribed time. The resulting dishes can be prepared with greater convenience as no pots or pans are dirtied in the process. Such meals are available for camping as well as home dining.

Heating

Heat treatment or thermal processing of food is used in order to kill or inactivate bacteria, increase the shelf-life, or create products with an attractive appearance. Today's cooled displays contain healthy and palatable ready-to-eat meals several weeks after they were produced. This is possible because the food has undergone heat treatment during production.

Minimal Thermal Processing

The food is heated to a low temperature for a short period and acquires a fresh and attractive appearance, while retaining most of the nutrients. The technique is particularly suitable for food that does not need a shelf-life of longer than 7-12 days. Catering establishments for canteens, sheltered accommodation and hospitals often use this technique, and it is used also by restaurants and other food outlets.

Cook-chill

The cook-chill method is used to preserve as much flavour and moisture as possible in the products. The food is heat treated at a temperature and time depending on raw material and desired product quality. After rapid cooling the food is packaged and stored at chilled temperatures. The shelf-life of such products is generally limited to 10-14 days, depending on the storage conditions and raw material. The method is often combined with modified atmosphere packaging.

Sous Vide

Sous vide is used to describe food that has been vacuum-packed and given mild heat treatment before being stored in at chilled conditions. The heat treatment takes place after packaging and under controlled conditions of time and temperature, and the food is subsequently rapidly cooled. The products are kept chilled until they are heated before serving. Sous vide is now used with a more intense heat treatment, and this has made it possible to produce ready-to-eat food that has a shelf life of several weeks. Testing the nutritional value of four ready-to-eat meals has shown the method to be just as good as corresponding home-cooking.

Pickling

The term pickle is derived from the Dutch word pekel, meaning brine. Pickling, also known as brining or corning is the process of preserving food by anaerobic fermentation in brine (a solution of salt in water) to produce lactic acid, or marinating and storing it in an acid solution, usually vinegar (acetic acid). The resulting food is called a pickle. This procedure gives the food a salty or sour taste. In South Asia, edible oils are used as the pickling medium with vinegar. Another distinguishing characteristic is a pH less than 4.6, which is sufficient to kill most bacteria. Pickling can preserve perishable foods for months. Antimicrobial herbs and spices, such as mustard seed, garlic, cinnamon or cloves, are often added. If the food contains sufficient moisture, a pickling brine may be produced simply by adding dry salt. For example, sauerkraut and Korean kimchi are produced by salting the vegetables to draw out excess water. Natural fermentation at room temperature, by lactic acid bacteria, produces the required acidity. Other pickles are made by placing vegetables in vinegar. Unlike the canning process, pickling (which includes fermentation) does not require that the food be completely sterile before it is sealed. The acidity or salinity of the solution, the temperature of fermentation, and the exclusion of oxygen determine which microorganisms dominate, and determine the flavor of the end product When both salt concentration and temperature are low, Leuconostocmesenteroides dominates, producing a mix of acids, alcohol, and aroma compounds. At higher temperatures Lactobacillus plantarum dominates, which produces primarily lactic acid. Many pickles start with Leuconostoc, and change to Lactobacillus with higher acidity. Pickling began as a way to preserve food for out-of-season use and for long journeys, especially by sea. Salt pork and salt beef were common staples for sailors before the days of steam engines. Although the process was invented to preserve foods, pickles are also made and eaten because people enjoy the resulting flavors. Pickling may also improve the nutritious value of food by introducing B vitamins produced by bacteria.

The exact origins of pickling are unknown, but the ancient Mesopotamians may have used the process around 2400 B.C. Pickling was used as a way to preserve food for out-of-season use and for long journeys, especially by sea. Salt pork and salt beef were common staples for sailors before the days of steam engines. Although the process was invented to preserve foods, pickles are also made and eaten because people enjoy the resulting flavors. Pickling may also improve the nutritional value of food by introducing B vitamins produced by bacteria.

The term pickle is derived from the Dutch word pekel, meaning brine. In the U.S. and Canada, and sometimes Australia and New Zealand, the word pickle alone almost always refers to a pickled cucumber, except when it is used figuratively. It may also refer to other types of pickles such as "pickled onion", "pickled cauliflower", etc. In the UK, pickle, as in a "cheese and pickle sandwich", may also refer to Ploughman's pickle, a kind of chutney.

Process of Pickling

Pickling is a method of preserving food in an edible anti-microbial liquid. Pickling is of two types:

- Chemical pickling (brining).
- Fermentation pickling.

In chemical pickling, food is placed in an edible liquid that inhibits or kills bacteria and other micro-organisms. A number of pickling agents can be used like brine, vinegar, alcohol, and vegetable oil (olive or mustard oil). The chemical pickling process could also involve heating or boiling so that the food being preserved becomes saturated with the pickling agent. Common chemically pickled foods include cucumbers, peppers, corned beef, herring, and eggs, chow-chow (chayote and Hindi *chocho)* as well mixed vegetables such as piccalilli and giardiniera (chopped pickled vegetables and spices) and also Indian achar.

In fermented pickling, the food itself produces the preservation agent, typically by a process that produces lactic acid. Fermented pickles include sauerkraut, nukazuke (Japanese pickle made by fermenting vegetables in rice bran) kimchi, surströmming (Swedish food- fermented Baltic Sea herring) and curtido (fermented cabbage, onions, carrots, oregano- Central American cuisine). Some chemically pickled cucumbers are also fermented. In commercial pickles, a preservative like sodium benzoate or EDTA may also be added to increase shelf life.

Popularity of Pickles Around the World

South Asia

South Asia has a large variety of pickles (known as *achar* in Assamese, Bengali, Hindi, Punjabi, *uppinakaayi* in Kannada, *lonacha* in Marathi, *uppilittathu* or *achar* in Malayalam, *oorukai* in Tamil, *ooragaya* in Telugu), which are mainly made from varieties of mango, lemon, lime, goongura(a sour leafy shrub), tamarind and Indian gooseberry (amla), chilli. Vegetables such as eggplant, carrots, cauliflower, tomato, bitter gourd, green tamarind, ginger, garlic, onion, and citron are also occasionally used. These fruits and vegetables are generally mixed with ingredients like salt, spices, and vegetable oils and are set to mature in a moistureless medium.

In Pakistan, pickles are known locally as *achaar* (in Urdu) and come in a variety of flavors. A popular item is the traditional mixed Hyderabadi pickle, a common delicacy prepared from an assortment of fruits (most notably mangoes) and vegetables blended with selected spices.

In Sri Lanka, *achcharu* is traditionally prepared from carrots, onions, and ground dates that are mixed with mustard powder, ground pepper, crushed ginger, garlic, and vinegar, and left to sit in a clay pot.

Southeast Asia

Singapore, Indonesian and Malaysian pickles, called *acar*, are typically made out of cucumber, carrot, bird's eye chilies, and shallots, these items being seasoned with vinegar, sugar and salt. Fruits, such as papaya and pineapple, are also sometimes pickled.

In the Philippines, *achara* is primarily made out of green papaya, carrots, and shallots, with cloves of garlic and vinegar. Other versions could include ginger, bell peppers, white radishes, cucumbers or bamboo shoots. Separately, in some provinces, unripe mangoes or *burong mangga*, unripe tomatoes, guavas, jicama, bitter gourd and other fruit and vegetables are also pickled. Siling labuyo, sometimes with garlic and red onions, are also pickled in bottled vinegar. The spiced vinegar itself is a staple condiment in Filipino cuisine.

In Vietnam, vegetable pickles are called *dưa muối* ("salted vegetables") or *dưa chua* ("sour vegetables"). In Burma, tea leaves are pickled to produce lahpet, which has strong social and cultural importance.

Kimchi is a very common side dish in Korea.

East Asia

China is home to a huge variety of pickled vegetables, including radish, *baicai* (Chinese cabbage, notably *suan cai, la bai cai, pao cai*, and Tianjin preserved vegetable), *zha cai*, chili pepper, and cucumbers, among many others.

Japanese *tsukemono* (pickled foods) include *takuan* (daikon), *umeboshi* (ume plum), *gari* & *beni shoga* (ginger), turnip, cucumber, and Chinese cabbage.

The Korean staple kimchi is usually made from pickled napa cabbage and radish, but is also made from green onions, garlic stems, chives and a host of other vegetables. Kimchi is popular throughout East Asia. Jangajji is another example of pickled vegetables.

Western Asia

Torshi, traditional pickles in Southeast Europe, Western Asia and the Caucasus.

In Iran, Turkey, Arab countries, the Balkans, and the Caucasus, pickles (called torshi in Persian, *turşu* in Turkish language and *mekhallel* in Arabic) are commonly made from turnips, peppers, carrots, green olives, cucumbers, cabbage, green tomatoes, lemons, and cauliflower.

Central and Eastern Europe

Coriander seeds are one of the spices popularly added to pickled vegetables in Europe.

In Hungary the main meal *(lunch)* usually goes with some kind of pickles *(savanyúság)* but they are commonly consumed at other times of the day too. The most commonly consumed pickles are sauerkraut *(savanyú káposzta)*, the different kinds of pickled cucumbers and peppers and *csalamádé* but tomatoes, carrots, beetroot, baby corn, onions, garlic, certain squashes and melons and a few fruits like plums and apples are used to make pickles too. Stuffed pickles are specialties usually made of peppers or melons pickled after being stuffed with a cabbage filling. Pickled plum stuffed with garlic is a unique

Hungarian type of pickle just like *csalamádé* and leavened cucumber *(kovászos uborka)*. *Csalamádé* a type of mixed pickle made of cabbage, cucumber, paprika, onion, carrot, tomatoes and bay leaf mixed up with vinegar as the fermenting agent. Leavened cucumber, unlike other types of pickled cucumbers that are around all year long, is rather a seasonal pickle produced in the summer. Cucumbers, spices, herbs and slices of bread are put in a glass jar with salt water and kept in direct sunlight for a few days. The yeast from the bread, along with other pickling agents and spices fermented under the hot sun, give the cucumbers a unique flavor, texture and slight carbonation. Its juice can be used to make a special type of spritzer *('Újházy fröccs')* instead of carbonated water. It is common for Hungarian households to produce their own pickles. Different regions or towns have their special recipes unique to them. Among them all the Vecsési Sauerkraut *(Vecsési savanyú káposzta)* is the most famous.

Pickled tomatoes are common in CIS.

Romanian pickles (murături) are made out of beetroot, cucumbers, green tomatoes (*gogonele*), carrots, cabbage, garlic, sauerkraut (bell peppers stuffed with cabbage), bell peppers, melons, mushrooms, turnips, celery and cauliflower. Meat, like pork, can also be preserved in salt and lard.

Polish, Czech and Slovak traditional pickles are cucumbers and sauerkraut, but other pickled fruits and vegetables, including plums, pumpkins and mushrooms are also common.

Russian, Ukrainian and Belarusian pickled items include beets, mushrooms, tomatoes, sauerkraut, cucumbers, ramsons, garlic, eggplant (which is typically stuffed with julienned carrots), custard squash, and watermelon. Garden produce is commonly pickled using salt, dill, blackcurrant leaves, bay leaves and garlic and is stored in a cool, dark place. The leftover brine (called *rassol* (рассол) in Russian) has a number of culinary uses in these countries, especially for cooking traditional soups, such as shchi, rassolnik, and solyanka. *Rassol*, especially cucumber or sauerkraut rassol, is also a favorite traditional remedy against morning hangover.

Southern Europe

An Italian pickled vegetable dish is giardiniera, which includes onions, carrots, celery and cauliflower. Many places in southern Italy, particularly in Sicily, pickle eggplants and hot peppers.

In Albania, Bulgaria, Serbia, Macedonia and Turkey, mixed pickles, known as *turshi*, *tursija* or *turshu* form popular appetizers, which are typically eaten with *rakia*. Pickled green tomatoes, cucumbers, carrots, bell peppers, peppers, eggplants, and sauerkraut are also popular.

Turkish pickles, called *turşu*, are made out of vegetables, roots, and fruits such as peppers, cucumber, Armenian cucumber, cabbage, tomato, eggplant (aubergine), carrot, turnip, beetroot, green almond, baby watermelon, baby cantaloupe, garlic, cauliflower, bean and green plum. A mixture of spices flavor the pickles.

In Greece, pickles, called *τουρσί(α)*, are made out of carrots, celery, eggplants stuffed with diced carrots, cauliflower, tomatoes, and peppers.

Northern Europe

In Britain, pickled onions and pickled eggs are often sold in pubs and fish and chip shops. Pickled beetroot, walnuts, and gherkins, and condiments such as Branston Pickle and piccalilli are typically eaten as an accompaniment to pork pies and cold meats, sandwiches or a ploughman's lunch. Other popular pickles in the UK are pickled mussels, cockles, red cabbage, mango chutney, sauerkraut, and olives. Rollmops are also quite widely available under a range of names from various producers both within and out of the UK.

Pickled herring, rollmops, and salmon are popular in Scandinavia. Pickled cucumbers and red garden beets are important as condiments for several traditional dishes. Pickled capers are also common in Scandinavian cuisine.

United States and Canada

A dish of giardiniera.

In the United States and Canada, pickled cucumbers (most often referred to simply as "pickles" in Canada and the United States), olives, and sauerkraut are most commonly seen, although pickles common in other nations are also available.

Canadian pickling is similar to that of Britain. Through the winter, pickling is an important method of food preservation. Pickled cucumbers, onions, and eggs are common individual pickled foods seen in Canada. Chow-chow is a tart vegetable mix popular in the Maritime Provinces and the Southern United States, similar to piccalilli. Pickled fish is commonly seen, as in Scotland. Meat is often also pickled or preserved in different brines throughout the winter, most prominently in the harsh climate of Newfoundland.

Giardiniera, a mixture of pickled peppers, celery and olives, is a popular condiment in Chicago and other cities with large Italian-American populations, and is often consumed with Italian beef sandwiches. Pickled eggs are common in the Upper Peninsula of Michigan. Pickled herring is available in the Upper Midwest. Pennsylvania Dutch Country has a strong tradition of pickled foods, including chow-chow and red beet eggs. In the Southern United States, pickled okra and watermelon rind are popular, as are deep-fried pickles and pickled pig's feet, pickled chicken eggs, pickled quail eggs, pickled garden vegetables and pickled sausage. In Mexico, chili peppers, particularly of the Jalapeño and serrano varieties, pickled with onions, carrots and herbs form common condiments. Various pickled vegetables, fish, or eggs may make a side dish to a Canadian lunch or dinner. Popular pickles in the Pacific Northwest include pickled asparagus and green beans. Pickled fruits like blueberries and early green strawberries are paired with meat dishes in restaurants.

In the United States, National Pickle Day is recognized as a food "holiday" every year on November 14.

Mexico, Central America and South America

In the Mesoamerican region pickling is known as "encurtido" or "curtido" for short. The pickles or "curtidos" as known in Latin America are served cold, as an appetizer, as a side dish or as a tapas dish in Spain. In several Central American countries it is prepared with cabbage, onions, carrots, lemon, vinegar, oregano, and salt. In Mexico, "curtido" consists of carrots, onions, and jalapeño peppers and used to accompany meals still common in taquerias and restaurants. In order to prepare a carrot "curtido" simply add carrots to vinegar and other ingredients that are common to the region such as chilli, tomato, and onions. Varies depending on the food, in the case of sour. Another example of a type of pickling which involves the pickling of meats or seafood is the "escabeche" or "ceviches" popular in Peru, Ecuador, and throughout Latin America and the Caribbean. These dishes include the pickling of pig's feet, pig's ears, and gizzards prepared as an "escabeche" with spices and seasonings to flavor it. The ceviches consists of shrimp, octopus, and various fishes seasoned and served cold.

Possible Health Hazards of Pickled Vegetables

The World Health Organization has listed pickled vegetables as a possible carcinogen, and the *British Journal of Cancer* released an online 2009 meta-analysis of research on pickles as increasing the risks of esophageal cancer. The report, citing limited data in a statistical meta analysis, indicates a potential two-fold increased risk of oesophageal cancer associated with Asian pickled vegetable consumption. Results from the research are described as having "high heterogeneity" and the study said that further well-designed prospective studies were warranted. However, their results stated "The majority of subgroup analyses showed a statistically significant association between consuming pickled vegetables and Oesophageal Squamous Cell Carcinoma".

The 2009 meta-analysis reported heavy infestation of pickled vegetables with fungi. Some common fungi can facilitate the formation of N-nitroso compounds, which are strong oesophageal carcinogens in several animal models. Roussin red methyl ester, a non-alkylating nitroso compound with tumour-promoting effect in vitro, was identified in pickles from Linxian in much higher concentrations than in samples from low-incidence areas. Fumonisin mycotoxins have been shown to cause liver and kidney tumours in rodents.

A 2017 study in *Chinese Journal of Cancer* has linked salted vegetables (common among Chinese cuisine) to a 4-fold increase in nasopharynx cancer, where fermentation was a critical step in creating nitrosamines, which some are confirmed carcinogens, as well as activation of Epstein–Barr virus by fermentation products.

Pickled cucumbers

pickled herring Pickled

mushrooms

Pickled olives

Pickled vegetables

Fermented homemade pickled cucumber, chili pepper, garlic, and apple in the hot climate of Indonesia.

Sugaring

Sugaring is a method of food preservation that requires the food to be dehydrated and then to be packed with either crystallized sugar or with the liquids containing high amount of sugar such as honey or molasses.

The main purpose of this food preservation method is to treat the food in order to stop the growth of bacteria that may diminish the nutritional value and quality of the food. Some of the most popular sugared foods that are found in almost every cuisine across the globe include different kinds of fruits as well as a variety of vegetables such as ginger.

Although there is no specific evidence present to prove the origin of the sugaring as a food preservation method, the methods of preserving food adopted by the early American colonists used to include sugaring as one of the methods. Unlike today, early Americans were not so well-equipped with the techniques and means of transporting food from different climates and they could only rely upon the seasonal fruits and vegetables. With the lack of refrigeration, they were not able to store food for longer periods, therefore they developed various food preserving procedures that helped them in retaining the nutritional value and quality of food for extended period of time.

All the food preserving techniques had been introduced through trial and error, but the actual working of the methods was not known as bacteria had not been known till then. Though sugaring food preservation is believed to be done by the early Americans, the versatility of this food preservation made it popular in all cuisines.

The Process of Sugaring

Sugaring mainly reduces the action of the internal food enzymes by inhibiting the bacterial growth and this result in the preservation of food from spoiling.

The process of this type of food preservation is mainly done with any table or raw sugar that is generally in crystallized form. The sugar can also be liquefied in form of syrup, honey or molasses. These high sugar density liquids are of equal importance as sugar in the process of food preservation. Any fruit or vegetable that has to be preserved is washed thoroughly and desiccated by dehydration. The food is then cooked in liquid sugar product or raw sugar until crystallized and the resultant food is preserved in dried form.

Some fruits are even glazed in sugar syrup, but sold after being extracted from the liquid. The sugary coating on the fruits and the internal high sugar content increases the shelf life of the food.

Alcoholic preservation combined with sugar is also extremely popular to preserve some luxury items such as brandy and other spirits containing fruits. This type of food preservation should not be confused with the fruit flavored spirits like cherry brandy or apple wine.

Sugaring of Various Foods:

- Fruits – Apples, pears, plum, cherry, apricots and peaches are some of the popular fruits that are commonly preserved by sugaring method. These fruits are either dried before preservation or glazed in sugar syrup. Fruits are also combined with alcohol and preserved.
- Vegetables – Ginger and carrot are the most common vegetables that are often sugared to prepare relishes or sweet pickles. These candied vegetables are popularly served as condiments.
- Angelica – It is a herb that is widely used as a flavoring agent. However, sugar preserved or candied strips of angelica are extremely popular as cake decorations.

- Citrus peel – The peels of citrus fruits like lemon and amla (Indian gooseberry) are often candied to form relishes. However, 'murabba' (Indian candied dish) includes whole amla.

Advantages and Disadvantages of Sugaring Food Preservation Method

Sugaring has few advantages over other preservation methods, as this process does not require large number of ingredients and often the sugar extract or glaze is used to sweeten various other foods as well. It is also an easy preservation method with less time involvement. There is a risk in this method as sugar is believed to attract moisture very fast. When the atmospheric moisture is high in content, the yeast present in the environment starts its action and sugar starts fermenting into carbon-di-oxide and alcohol. Although fermented food is also a preserved food, the sugared foods should be prevented from fermenting, as it may lead to an unpleasant taste.

Canning

Canning is the method of preserving food from spoilage by storing it in containers that are hermetically sealed and then sterilized by heat. The process was invented after prolonged research by Nicolas Appert of France in 1809, in response to a call by his government for a means of preserving food for army and navy use. Appert's method consisted of tightly sealing food inside a bottle or jar, heating it to a certain temperature, and maintaining the heat for a certain period, after which the container was kept sealed until use. It was 50 years before Louis Pasteur was able to explain why the food so treated did not spoil: the heat killed the microorganisms in the food, and the sealing kept other microorganisms from entering the jar. In 1810 Peter Durand of England patented the use of tin-coated iron cans instead of bottles, and by 1820 he was supplying canned food to the Royal Navy in large quantities. European canning methods reached the United States soon thereafter, and that country eventually became the world leader in both automated canning processes and total can production. In the late 19th century, Samuel C. Prescott and William Underwood of the United States set canning on a scientific basis by describing specific time-temperature heating requirements for sterilizing canned foods.

Originally, cans consisted of a sheet of tin-plated iron that was rolled into a cylinder (known as the body), onto which the top and bottom were manually soldered. This form was replaced in the early

20th century by the modern sanitary, or open-top, can, whose constituent parts are joined by interlocking folds that are crimped, or pressed together. Polymer sealing compounds are applied to the end, or lid, seams, and the body seams can be sealed on the outside by soldering. The modern tin can is made of 98.5 percent sheet steel with a thin coating of tin (i.e., tinplate). It is manufactured on wholly automatic lines of machinery at rates of hundreds of cans per minute.

Most vegetables, fruits, meat and dairy products, and processed foods are stored in tin cans, but soft drinks and many other beverages are now commonly stored in aluminum cans, which are lighter and do not rust. Aluminum cans are made by impact extrusion; the body of the can is punched out in one piece from a single aluminum sheet by a stamping die. This seamless piece, which has a rounded bottom, is then capped with a second piece as its lid. The tabs used in pop-top cans are also made of aluminum. Bimetal cans are made of aluminum bodies and steel lids.

Canneries are usually located close to the growing areas of the product to be packed, since it is desirable to can foods as quickly as possible after harvesting. The canning process itself consists of several stages: cleaning and further preparing the raw food material; blanching it; filling the containers, usually under a vacuum; closing and sealing the containers; sterilizing the canned products; and labeling and warehousing the finished goods. Cleaning usually involves passing the raw food through tanks of water or under high-pressure water sprays, after which vegetable or other products are cut, peeled, cored, sliced, graded, soaked, pureed, and so on. Almost all vegetables and some fruits require blanching by immersion in hot water or steam; this process softens the vegetable tissues and makes them pliable enough to be packed tightly, while also serving to inactivate enzymes that can cause undesirable changes in the food before canning. Blanching also serves as an additional or final cleansing operation.

The filling of cans is done automatically by machines; cans are filled with solid contents and, in many cases, with an accompanying liquid (often brine or syrup) in order to replace as much of the air in the can as possible. The filled cans are then passed through a hot-water or steam bath in an exhaust box; this heating expands the food and drives out the remaining air; thus, after sealing, heat sterilizing, and cooling the can, the contraction of the contents produces a partial vacuum within the container. Certain products are vacuum-packed, whereby the cans are mechanically exhausted by specially designed vacuum-can sealing machines.

Immediately after the cans are exhausted, they are closed and sealed; a machine places the cover on the can, and the curl on the can cover and the flange on the can body are rolled into position and then flattened together. The thin layer of sealing compound originally present in the rim of the cover is dispersed between the layers of metal to ensure a hermetic seal. The sealed cans are then sterilized; i.e., they are heated at temperatures high enough and for a long enough time to destroy all microorganisms (bacteria, molds, yeasts) that might still be present in the food contents. The heating is done in high-pressure steam kettles, or cookers, usually using temperatures around 240°F (116°C). The cans are then cooled in cold water or air, after which they are labeled.

Canning preserves most of the nutrients in foods. Proteins, carbohydrates, and fats are unaffected, as are vitamins A, C, D, and B2. The retention of vitamin B1 depends on the amount of heat used during canning. Some vitamins and minerals may dissolve into the brine or syrup in a can during processing, but they retain their nutritive value if those liquids are consumed.

There are two methods of canning, i.e., water-bath canning and pressure canning:

- Water-bath canning: This method, sometimes referred to as *hot water canning,* uses a large kettle of boiling water. Filled jars are submerged in the water and heated to an internal temperature of 212 degrees for a specific period of time. Use this method for processing high-acid foods, such as fruit, items made from fruit, pickles, pickled food, and tomatoes.

- Pressure canning: Pressure canning uses a large kettle that produces steam in a locked compartment. The filled jars in the kettle reach an internal temperature of 240 degrees under a specific pressure (stated in pounds) that's measured with a dial gauge or weighted gauge on the pressure-canner cover. Use a pressure canner for processing vegetables and other low-acid foods, such as meat, poultry, and fish.

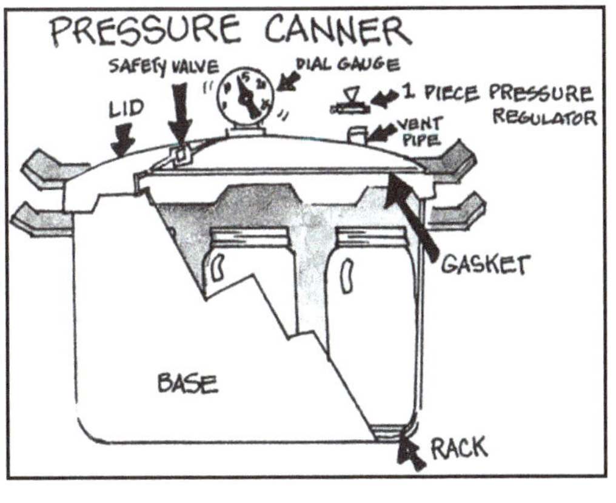

Jellying

Food may be preserved by cooking in a material that solidifies to form a gel. Such materials include gelatin, agar, maize flour, and arrowroot flour. Some foods naturally form a protein gel when cooked, such as eels and elvers, and sipunculid worms, which are a delicacy in Xiamen, in the Fujian province of the People's Republic of China. Jellied eels are a delicacy in the East End of London, where they are eaten with mashed potatoes. Potted meats in aspic (a gel made from gelatin and clarified meat broth) were a common way of serving meat off-cuts in the UK until the 1950s. Many jugged meats are also jellied.

A traditional British way of preserving meat (particularly shrimp) is by setting it in a pot and sealing it with a layer of fat. Also common is potted chicken liver; jellying is one of the steps in producing traditional pates.

Burial

Burial of food can preserve it due to a variety of factors: lack of light, lack of oxygen, cool temperatures, pH level, or desiccants in the soil. Burial may be combined with other methods such as salting or fermentation. Most foods can be preserved in soil that is very dry and salty (thus a desiccant) such as sand, or soil that is frozen. Many root vegetables are very resistant to spoilage and require no other preservation than storage in cool dark conditions, for example by burial in the ground, such as in a storage clamp. Century eggs are created by placing eggs in alkaline mud (or other alkaline substance), resulting in their "inorganic" fermentation through raised pH instead of spoiling. The fermentation preserves them and breaks down some of the complex, less flavorful proteins and fats into simpler, more flavorful ones. Cabbage was traditionally buried in the fall in northern farms in the U.S. for preservation. Some methods keep it crispy while other methods produce sauerkraut. A similar process is used in the traditional production of kimchi. Sometimes meat is buried under conditions that cause preservation. If buried on hot coals or ashes, the heat can kill pathogens, the dry ash can desiccate, and the earth can block oxygen and further contamination. If buried where the earth is very cold, the earth acts like a refrigerator.

Fermentation

Fermentation in food processing typically is the conversion of carbohydrates to alcohols and carbon dioxide or organic acids using yeasts, bacteria, or a combination thereof, under anaerobic conditions. Fermentation in simple terms is the chemical conversion of sugars into ethanol. The science of fermentation is also known as zymology or zymurgy.

Historically, when studying the fermentation of sugar to alcohol by yeast, Louis Pasteur concluded that the fermentation was catalyzed by a vital force, called "ferments," within the yeast cells. The "ferments" were thought to function only within living organisms. "Alcoholic fermentation is an act correlated with the life and organization of the yeast cells, not with the death or putrefaction of the cells," he wrote.

Fermentation usually implies that the action of microorganisms is desirable. This process is used to produce alcoholic beverages such as wine, beer, and cider. Fermentation is also employed in the leavening of bread (CO_2 produced by yeast activity); in preservation techniques to produce lactic acid in sour foods such as sauerkraut, dry sausages, kimchi, and yogurt; and in the pickling of foods with vinegar (acetic acid).

Beer: This is beer fermenting at a brewery.

Food fermentation has been said to serve five main purposes:

1. Enrichment of the diet through development of a diversity of flavors, aromas, and textures in food substrates.
2. Preservation of substantial amounts of food through lactic acid, alcohol, acetic acid, and alkaline fermentations.
3. Biological enrichment of food substrates with protein, essential amino acids, essential fatty acids, and vitamins.
4. Elimination of antinutrients.
5. A decrease in cooking time and fuel requirement.

Process of Fermentation Preserves Foods

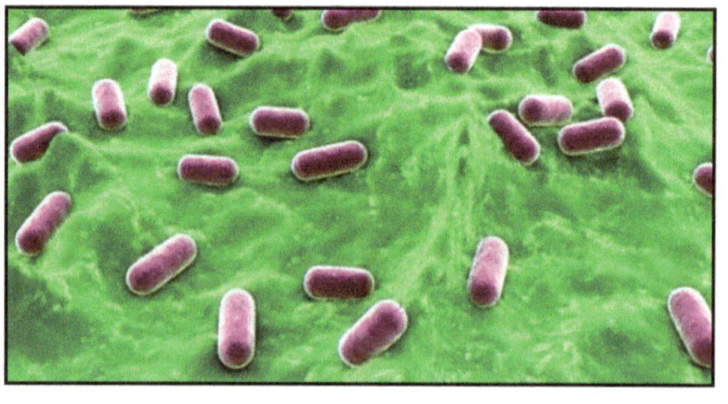

The desirable bacteria cause less deterioration of the food by inhibiting the growth of the spoiling types of bacteria. Some fermenting processes lower the pH of foods preventing harmful microorganisms to live with too acidic an environment. Controlled fermentation processes encourage the growth of good bacteria which starves, or fights off, the bad microbes. Depending on what is fermented, or the manner of fermentation, foods can remain consumable for years.

The fermentation process can be stopped by other means of preserving, such as, canning (heating), drying, or freezing. Heat (pasteurization, 145°F), and low temperatures (freezing, 32°F or below) stops the fermenting process by slowing, or killing, the preferred microorganisms, and other bacteria. A few undesirable bacteria are not killed by either means, and continue to grow. When the beneficial bacteria are gone, the unfavorable bacteria take over, growing exponentially! This causes rotting, disease, illness, and inedible foods. When the good guys are present and happy, the food remains edible.

Additional Benefits of Fermenting

Fermenting enhances the flavors of some foods, as with the extended fermentation of black teas, aged cheese, wine, and beer, which creates their distinctive flavors. Cocoa beans have to be fermented (composted) for a few days to remove the pods and to enhance the flavor of chocolate.

Fermenting makes foods more edible by changing chemical compounds, or predigesting, the foods for us. There are extreme examples of poisonous plants like cassava that are converted to edible products by fermenting. Some coffee beans are hulled by a wet fermenting process, as opposed to a dry process.

Fermentation increases nutritional values with the biochemical exchange it produces, and allows us to live healthier lives. Here are a few examples:

- The sprouting of grains, seeds, and nuts, multiplies the amino acid, vitamin, and mineral content and antioxidant qualities of the starting product.
- Fermented beans are easier for our bodies to digest, like the proteins found in soy beans that are nearly indigestible until fermented.
- Fermented dairy products, like, cheese, yogurt, and kifir, can be consumed by those not able to digest the raw milk, and aid the digestion and well-being for those with lactose intolerance and autism.
- Porridge made from grains allowed to ferment increases the nutritional values so much that it reduces the risk of disease in children.
- The news is full of reports about the health benefits of probotic supplements (beneficial bacterial cultures for microbial balance in the body) fighting cancer and other diseases.
- Vinegar is used to leach out certain flavors and compounds from plant materials to make healthy and tasty additions to our meals.

Pasteurization

Pasteurization is the process of heat processing a liquid or a food to kill pathogenic bacteria to make the food safe to eat. The use of pasteurization to kill pathogenic bacteria has helped reduce the transmission of diseases, such as typhoid fever, tuberculosis, scarlet fever, polio, and dysentery.

It is important to note that foods can become contaminated even after they have been pasteurized. For example, all pasteurized foods must be refrigerated. If the pasteurized food is temperature-abused (e.g., if milk or eggs are not kept refrigerated), it could become contaminated. Therefore, it is important to *always* handle food properly by handling it with clean hands, preventing it from becoming contaminated, and keeping it at a safe temperature.

Process of Pasteurization

Foods are heat-processed to kill pathogenic bacteria. Foods can also be pasteurized using gamma irradiation. Such treatments do not make the foods radioactive. The pasteurization process is based on the use of one of following time and temperature relationships.

High-Temperature-Short-Time Treatment (HTST) -- this process uses higher heat for less time to kill pathogenic bacteria. For example, milk is pasteurized at 161°F (72°C) for 15 seconds.

Low-Temperature-Long-Time Treatment (LTLT) -- this process uses lower heat for a longer time to kill pathogenic bacteria. For example, milk is pasteurized at 145°F (63°C) for 30 minutes.

It is important to remember that the times and temperatures depend on: (1) the type of food and (2) the final result one wants to achieve, such as retaining a food's nutrients, color, texture, and flavor.

Processes used to Pasteurize Foods

Flash Pasteurization - Involves a high-temperature, short-time treatment in which pourable products, such as juices, are heated for 3 to 15 seconds to a temperature that destroys harmful micro-organisms. After heating, the product is cooled and packaged. Most drink boxes and pouches use this pasteurization method as it allows extended unrefrigerated storage while providing a safe product.

Steam Pasteurization - This technology uses heat to control or reduce harmful microorganisms in beef. This system passes freshly-slaughtered beef carcasses that are already inspected, washed, and trimmed, through a chamber that exposes the beef to pressurized steam for approximately 6 to 8 seconds. The steam raises the surface temperature of the carcasses to 190° to 200°F (88° to 93°C). The carcasses are then cooled with a cold-water spray. This process has proven to be successful in reducing pathogenic bacteria, such as *E. coli* O157:H7, *Salmonella*, and *Listeria*, without the use of any chemicals. Steam pasteurization is used on nearly 50% of U.S. beef.

Irradiation Pasteurization - Foods, such as poultry, red meat, spices, and fruits and vegetables, are subjected to small amounts of gamma rays. This process effectively controls vegetative bacteria and parasitic foodborne pathogens and increases the storage time of foods.

The Effect of Pasteurization on Nutrients and Flavor

Pasteurization can affect the nutrient composition and flavor of foods. In the case of milk, for example, the high-temperature, short-time treatments (HTST) cause less damage to the nutrient composition and sensory characteristics of foods than do the low-temperature, long-time treatments (LTLT).

Examples of Foods that are Commonly Pasteurized

Whole Eggs Removed from Shells and Sold as a Liquid - Large quantities of eggs are sold to restaurants and institutions out of the shell. The yolk and whole-egg products are pasteurized in their raw form. The egg white is pasteurized in its raw form if it is sold as a liquid or frozen product.

Dried Eggs - If eggs are sold dried, the egg white with the glucose removed is normally heat-treated in the container by holding it for 7 days in a hot room at a minimum temperature of 130°F (54°C).

Whole Eggs Pasteurized in the Shell - Traditionally, eggs sold to customers in the shell have not been pasteurized. However, new time/temperature pasteurization methods are making this possible. Egg whites coagulate at 140°F (60°C). Therefore, heating an egg above 140°F would cook the egg, so processors pasteurize the egg in the shell at a low temperature, 130°F (54°C), for a long time, 45 minutes. This new process is being used by some manufacturers, but it is not yet widely available. Pasteurizing eggs reduces the risk of contamination from pathogenic bacteria, such as *Salmonella*, which can cause severe illness and even death. Pasteurized eggs in the shell may be used in recipes calling for raw eggs, such as Caesar salad, hollandaise or be arnaise sauces, mayonnaise, egg nog, ice cream, and egg-fortified beverages that are not thoroughly cooked.

Milk - Pasteurization improves the quality of milk and milk products and gives them a longer shelf life by destroying undesirable enzymes and spoilage bacteria. For example, the liquid is heated to 145°F (63°C) for at least 30 minutes or at least 161°F (72°C) for 15 seconds.

Today, many foods, such as eggs, milk, juices, spices and ice cream,are pasteurized. Sometimes higher temperatures are applied for a shorter period of time. The temperatures and times are determined by what is necessary to destroy pathogenic bacteria and other more heat-resistant disease-causing microorganisms that may be found in milk. The liquid is then quickly cooled to 40°F (4°C). Other liquids, such as juices, are heat-processed in a similar manner. Temperatures and times vary, depending on the product and the target organism.

Other Types of Milk Pasteurization

Ultrapasteurization - This involves the heating of milk and cream to at least 280° F (138° C) for at least 2 seconds, but because of less stringent packaging, they must be refrigerated. The shelf life of milk is extended 60 to 90 days. After opening, spoilage times for ultrapasteurized products are similar to those of conventionally pasteurized products.

Ultra-High-Temperature (UHT) Pasteurization -This involves heating milk or cream to 280° to 302°F (138° to 150°C) for 1 or 2 seconds. The milk is then packaged in sterile, hermetically-sealed (airtight) containers and can be stored without refrigeration for up to 90 days. After opening, spoilage times for UHT products are similar to those of conventionally pasteurized products.

Microwave Volumetric Heating

Microwave heating refers to dielectric heating due to polarization effects at a selected frequency band in a nonconductor. Microwave heating in foods occurs due to coupling of electrical energy from an electromagnetic field in a microwave cavity with the food and its subsequent dissipation within food product. This results in a sharp increase in temperature within the product. Microwave energy is delivered at a molecular level through the molecular interaction with the electromagnetic field, in particular, through molecular friction resulting from dipole rotation of polar solvents and from the conductive migration of dissolved ions. Water in the food is the primary dipolar component responsible for the dielectric heating. In an alternating current electric field, the polarity of the field is varied at the rate of microwave frequency and molecules attempt to alignthemselves with the changing field. Heat is generated rapidly as a result of internal molecular friction.

The second major mechanism of heating with microwaves is through the polarization of ions as a result of the back and forth movement of the ionic molecules trying to align themselves with the oscillating electric field.

Advantages

i. Microwave penetrates inside the food materials and, therefore, cooking takes place throughout the whole volume of food internally, uniformly, and rapidly, which significantly reduces the processing time and energy.

ii. Since the heat transfer is fast, nutrients and vitamins contents, as well as flavor, sensory characteristics, and color of food are well preserved.

iii. Minimum fouling depositions, because of the elimination of the hot heat transfer surfaces, since the piping used is microwave transparent and remains relatively cooler than the product.

iv. High heating efficiency (80% or higher efficiency can be achieved).

v. Perfect geometry for clean-in-place (CIP) system.

vi. Suitable for heat-sensitive, high-viscous.

vii. Multiphase fluids.

Factors Affecting Microwave Heating

Some physical, thermal, and electrical properties determine the absorption of microwave energy and simultaneous heating behavior of food materials in microwave processing. These properties/ factors are briefly discussed below.

Frequency

For food application, only two frequencies are allocated for microwave heating (915 and 2450 MHz) and therefore, these frequencies are of special interest. The corresponding wavelengths of these frequencies are 0.328 and 0.122 m, respectively. The wavelength has special significance as most interactions between the energy and materials take place in that region and generate instantaneous heat due to molecular friction. Food constituents except moisture, lipids, and ash are relatively inert to prescribed microwave frequencies. In addition, frequency (or wavelength) dictates equipment components such as magnetron, waveguide, and to some extent heating volume.

Dielectric Properties

The electrical properties of materials in the context of microwave and radiofrequency heating are known as dielectric properties, which provide a measure of how food materials interact with electromagnetic energy. Biological materials may be viewed as non ideal capacitors in that they have the ability to store and dissipate electrical energy from an electromagnetic field and the properties can be expressed in terms of a complex notation. The complex notation is characterized by dielectric permittivity with a real component, dielectric constant, and an imaginary component, dielectric loss.

Moisture Content

The moisture content significantly affects the dielectric properties of the food product and consequently penetration depth of the microwaves. Uneven heating rate is observed in high-moisture foods because of low microwave penetration depth. Low-moisture foods will have more uniform heating rate because of the deeper microwave penetration. The initial moisture content of the product and the rate of moisture evaporation play important roles during microwave heating. The heating behavior of water is phase dependent (liquid water versus solid ice phase) and also depends on the available free water content.

Mass

A direct relationship exists between the mass and the amount of absorbed microwave power, which should be applied to achieve the desired heating. For a smaller mass, batch oven is suitable, while a larger through put would often be better in large capacity equipment with conveyor. Such equipment has the added advantage of providing greater heating uniformity by moving the product through the microwave field. Each microwave oven has a critical (minimum) sample mass for its efficient operation. It is usually around250 ml water load in a 1 kW oven. Below this level, significant amount of microwave power is not absorbed into the product, and at very low loads they may damage the magnetron.

Temperature

Microwave heating is significantly affected by the level of sample temperature. Dielectric properties may vary with temperature, depending upon the material. Both temperature and moisture content can change during heating and therefore, those may have a combined effect on the dielectric constant, dielectric loss factor, loss tangent, and subsequently on the heating behavior. The initial temperature of the food product being heated by microwaves should either be controlled or known, so that the microwave power can be adjusted to obtain uniform final temperatures. If the microwave oven is preset to increase the product temperature from 20°C to 80°C, it will practically reach target temperature of 95°C with an initial product temperature of 35°C. To compensate the effect of higher initial temperature, the power of MW oven should be reduced or a higher sample mass should be used or the product should be heated for a shorter duration.

Geometry and Location of Foods

The shape of the food product does play an important role in the distribution of heat within the product heated in a microwave oven. It affects the depth of microwave penetration, and the heating rate and uniformity. Irregular-shaped products are subjected to non uniform heating due to the difference in product thickness. The closer the size (thickness) is to the wavelength, the higher will be the center temperature. Smaller particulates require less heat than larger ones. In addition, the more regular the shape, the more uniform will be the heat distribution within the product. A food of a spherical or cylindrical shape heats more evenly than a square. A higher surface-to-volume ratio enhances the heating rate. Therefore, the heating rate for a sphere will be different from that of a cylinder with the same volume. Placement has the most significant effect.

Thermal Properties

The heating characteristics of foods are dependent to a greater or lesser extent on some thermal properties such as thermal conductivity, density, and heat capacity. Thermal conductivity of food plays a significant role in microwave heating. Materials with higher thermal conductivity dissipate heat faster than the ones with lower conductivity during microwave heating. Food with high thermal conductivity will take lesser time to attain uniform temperature during holding.

Application in Food Industry

The major industrial applications of microwave heating are tempering of frozen meat and poultry products; precooking of bacon for foodservice; sausage cooking; drying of various foods; baking of bread, biscuit, and confectionery; thawing of frozen products; blanching of vegetables; heating and sterilizing of fast food, cooked meals, and cereals; and pasteurization and sterilization of various foods.

Application in Dairy Industry

Milk is traditionally pasteurized in a heat exchanger before distribution. The application of microwave heating to pasteurize milk has been well studied and has been a commercial practice for quite a long time. The success of microwave heating of milk is based on established conditions that provide the desired degree of safety with minimum product quality degradation. Since the first reported study on the use of a microwave system for pasteurization of milk (Hamid et al 1969), several studies on microwave heating of milk have been carried out. The majority of these microwave-based studies have been used to investigate the possibility of shelf-life enhancement of pasteurized milk, application of microwave energy to inactivate milk pathogens, assess the influence on the milk nutrients or the non-uniform temperature distribution during the microwave treatment.

Irradiation

Food irradiation is a processing and preservation technique with similar results to freezing or pasteurisation. During this procedure, the food is exposed to doses of ionising energy, or radiation. At low doses, irradiation extends a product's shelf life. At higher doses, this process kills insects, moulds, bacteria and other potentially harmful micro-organisms.

Considerable scientific research over the past five decades indicates that food irradiation is a safe and effective form of processing. Food irradiation has been approved in 40 countries including Australia, the United States, Japan, China, France and Holland.

To date, in Australia and New Zealand, only herbs and spices, herbal infusions and some tropical fruits are approved for irradiation by Food Standards Australia New Zealand (FSANZ), in accordance with the FSANZ Food Standards Code. For each of these, FSANZ has established that there are no safety concerns and no significant nutritional changes to the food as a result of food irradiation. Irradiated foods will be clearly labelled so that consumers can make an informed choice.

Irradiated Foods and Radioactivity

There is a common misconception that irradiated food is radioactive. The radiation used to process foods is very different from the radioactive fallout that occurs after, for example, a nuclear accident.

In food processing, the radioactive sources permitted do not generate gamma, electrons or x-rays of sufficient high energy to make food radioactive. No radioactive energy remains in the food after treatment.

The World Health Organization (WHO), the American Dietetic Association and the Scientific Committee of the European Union are three internationally recognised bodies that support food irradiation.

Food Irradiation Procedure

The food is exposed to ionising radiation, either from gamma rays or a high-energy electron beam or powerful x-rays. Gamma rays and x-rays are a form of radiation that shares some characteristics with microwaves, but with much higher energy and penetration.

The rays pass through the food just like microwaves in a microwave oven, but the food does not heat up to any significant extent. Exposure to gamma rays does not make food radioactive. Electron beams and x-rays are produced using electricity, which can be switched on or off, and they do not require radioactive material.

In both cases, organisms that are responsible for spoiling foods– such as insects, moulds and bacteria, including some important food poisoning bacteria– can be killed. Food irradiation cannot kill viruses.

Effects of Irradiation on Food

Some foods, such as dairy foods and eggs, cannot be irradiated because it causes changes in flavour or texture. Fruits, vegetables, grain foods, spices and meats (such as chicken) can be irradiated.

Irradiation causes minimal changes to the chemical composition of the food, however, it can alter the nutrient content of some foods because it reduces the level of some of the B-group vitamins. This loss is similar to those that occur when food is cooked or preserved in more traditional and accepted ways, such as canning or blanching.

Labelling of Irradiated Foods

If a food has been irradiated or contains irradiated ingredients or components, it must be labelled with a statement that the food, ingredients or components have been treated with ionising radiation.

If a food product does not have a label (such as whole fruits sold loose), this statement must be displayed in close proximity to the food. In addition to mandatory labelling, irradiated foods may also be marked with a symbol called a 'radura', which is the international symbol for irradiation.

Proper Food Handling is Still Needed

Food irradiation can only be used if it fulfils a technological need or is necessary for a food safety or food hygiene purpose. It does not replace the need for correct food handling practices in industry and in the home. For instance, a few bacteria may survive the irradiation of meat. If the meat is left unrefrigerated, these bacteria could still multiply and cause food poisoning.

Uses

Irradiation is used to reduce or eliminate the risk of food-borne illnesses, prevent or slow down spoilage, arrest maturation or sprouting and as a treatment against pests. Depending on the dose, some or all of the pathogenic organisms, microorganisms, bacteria, and viruses present are destroyed, slowed down, or rendered incapable of reproduction. Irradiation cannot revert spoiled or over ripened food to a fresh state. If this food was processed by irradiation, further spoilage would cease and ripening would slow down, yet the irradiation would not destroy the toxins or repair the texture, color, or taste of the food. When targeting bacteria, most foods are irradiated to significantly reduce the number of active microbes, not to sterilize all microbes in the product. In this respect it is similar to pasteurization.

Irradiation is used to create safe foods for people at high risk of infection, or for conditions where food must be stored for long periods of time and proper storage conditions are not available. Foods that can tolerate irradiation at sufficient doses are treated to ensure that the product is completely sterilized. This is most commonly done with rations for astronauts, and special diets for hospital patients.

Irradiation is used to create shelf-stable products. Since irradiation reduces the populations of spoilage microorganisms, and because pre-packed food can be irradiated, the packaging prevents recontamination of the final product.

Irradiation is used to reduce post-harvest losses. It reduces populations of spoilage micro-organisms in the food and can slow down the speed at which enzymes change the food, and therefore slows spoilage and ripening, and inhibits sprouting (e.g., of potato, onion, and garlic).

Food is also irradiated to prevent the spread of invasive pest species through trade in fresh vegetables and fruits, either within countries, or trade across international boundaries. Pests such as insects could be transported to new habitats through trade in fresh produce which could significantly affect agricultural production and the environment were they to establish themselves. This "phytosanitary irradiation" aims to render any hitch-hiking pest incapable of breeding. The pests are sterilized when the food is treated by low doses of irradiation. In general, the higher doses required to destroy pests such as insects, mealybugs, mites, moths, and butterflies either affect the look or taste, or cannot be tolerated by fresh produce. Low dosage treatments (less than 1000 gray) enables trade across quarantine boundaries and may also help reduce spoilage.

Impact

Irradiation reduces the risk of infection and spoilage, does not make food radioactive, and the food is shown to be safe, but it does cause chemical reactions that alter the food and therefore alters the chemical makeup, nutritional content, and the sensory qualities of the food. Some of the potential secondary impacts of irradiation are hypothetical, while others are demonstrated. These effects include cumulative impacts to pathogens, people, and the environment due to the reduction of food quality, the transportation and storage of radioactive goods, and destruction of pathogens, changes in the way we relate to food and how irradiation changes the food production and shipping industries.

Immediate Effects

The radiation source supplies energetic particles or waves. As these waves/particles pass through a target material they collide with other particles. Around the sites of these collisions chemical bonds are broken, creating short lived radicals (e.g. the hydroxyl radical, the hydrogen atom and solvated electrons). These radicals cause further chemical changes by bonding with and or stripping particles from nearby molecules. When collisions damage DNA or RNA, effective reproduction becomes unlikely, also when collisions occur in cells, cell division is often suppressed.

Irradiation (within the accepted energy limits, as 10 MeV for electrons, 5 MeV for X-rays [US 7.5 MeV] and gamma rays from Cobalt-60) can not make food radioactive, but it does produce radiolytic products, and free radicals in the food. A few of these products are unique, but not considered dangerous.

Irradiation can also alter the nutritional content and flavor of foods, much like cooking. The scale of these chemical changes is not unique. Cooking, smoking, salting, and other less novel techniques, cause the food to be altered so drastically that its original nature is almost unrecognizable, and must be called by a different name. Storage of food also causes dramatic chemical changes, ones that eventually lead to deterioration and spoilage.

Misconceptions

A major concern is that irradiation might cause chemical changes that are harmful to the consumer. Several national expert groups and two international expert groups evaluated the available data and concluded that any food at any dose is wholesome and safe to consume as long as it remains palatable and maintains its technical properties (e.g. feel, texture, or color).

Irradiated food does not become radioactive, just as an object exposed to light does not start producing light. Radioactivity is the ability of a substance to emit high energy particles. When particles hit the target materials they may free other highly energetic particles. This ends shortly after the end of the exposure, much like objects stop reflecting light when the source is turned off and warm objects emit heat until they cool down but do not continue to produce their own heat. To modify a material so that it keeps emitting radiation (induce radiation) the atomic cores (nucleus) of the atoms in the target material must be modified.

It is impossible for food irradiators to induce radiation in a product. Irradiators emit electrons or photons and the radiation is intrinsically radiated at precisely known strengths (wavelengths for photons, and speeds for electrons). These radiated particles at these strengths can never be strong enough to modify the nucleus of the targeted atom in the food, regardless of how many particles hit the target material, and radioactivity can not be induced without modifying the nucleus.

Chemical Changes

Compounds known as free radicals form when food is irradiated. Most of these are oxidizers (i.e., accept electrons) and some react very strongly. According to the free-radical theory of aging excessive amounts of these free radicals can lead to cell injury and cell death, which may contribute to many diseases. However, this generally relates to the free radicals generated in the body, not the free radicals consumed by the individual, as much of these are destroyed in the digestive process.

When fatty acids are irradiated, a family of compounds called 2-alkylcyclobutanones (2-ACBs) are produced. These are thought to be unique radiolytic products. Most of the substances found in irradiated food are also found in food that has been subjected to other food processing treatments, and are therefore not unique. One family of chemicals (2ACB's) are uniquely formed by irradiation (unique radiolytic products), and this product is nontoxic. When irradiating food, all other chemicals occur in a lower or comparable frequency to other food processing techniques. Furthermore, the quantities in which they occur in irradiated food are lower or similar to the quantities formed in heat treatments.

The radiation doses to cause toxic changes are much higher than the doses used to during irradiation, and taking into account the presence of 2-ACBs along with what is known of free radicals, these results lead to the conclusion that there is no significant risk from radiolytic products.

Food Quality

Ionizing radiation can change food quality but in general very high levels of radiation treatment (many thousands of gray) are necessary to adversely change nutritional content, as well as the sensory qualities (taste, appearance, and texture). Irradiation to the doses used commercially to treat food have very little negative impact on the sensory qualities and nutrient content in foods. When irradiation is used to maintain food quality for a longer period of time (improve the shelf stability of some sensory qualities and nutrients) the improvement means that more consumers have access to the original taste, texture, appearance, and nutrients. The changes in quality and nutrition depend on the degree of treatment and may vary greatly from food to food.

There has been low level gamma irradiation that has been attempted on arugula, spinach, cauliflower, ash gourd, bamboo shoots, coriander, parsley, and watercress. There has been limited information, however, regarding the physical, chemical and/or bioactive properties and the shelf life on these minimally processed vegetables.

There is some degradation of vitamins caused by irradiation, but is similar to or even less than the loss caused by other processes that achieve the same result. Other processes like chilling, freezing, drying, and heating also result in some vitamin loss.

The changes in the flavor of fatty foods like meats, nuts and oils are sometimes noticeable, while the changes in lean products like fruits and vegetables are less so. Some studies by the irradiation industry show that for some properly treated fruits and vegetables irradiation is seen by consumers to improve the sensory qualities of the product compared to untreated fruits and vegetables.

Quality Impact on Minimally Processed Vegetables

Watercress (*Nasturtium Officinale*) is a rapidly growing aquatic or semi aquatic perennial plant. Because chemical agents do not provide efficient microbial reductions, watercress has been tested with gamma irradiation treatment in order to improve both safety and the shelf life of the product. It is traditionally used on horticultural products to prevent sprouting and post-packaging contamination, delay post-harvest ripening, maturation and senescence.

In a Food Chemistry food journal, scientists studied the suitability of gamma irradiation of 1, 2, and 5 kGy for preserving quality parameters of the fresh cut watercress at around 4 degrees Celsius for 7 days. They determined that a 2 kGy dose of irradiation was the dose that contained most similar qualities to non-stored control samples, which is one of the goals of irradiation. 2 kGy preserved high levels of reducing sugars and favoured PUFA; while samples of the 5 kGy dose revealed high contents of sucrose and MUFA. Both cases the watercress samples obtained healthier fatty acids profiles. However, a 5kGy dose better preserved the antioxidant activity and total flavonoids.

Long-term Impacts

If the majority of food was irradiated at high-enough levels to significantly decrease its nutritional content, there would be an increased risk of developing nutritionally-based illnesses if additional steps, such as changes in eating habits, were not taken to mitigate this. Furthermore, for at least three studies on cats, the consumption of irradiated food was associated with a loss of tissue in the myelin sheath, leading to reversible paralysis. Researchers suspect that reduced levels of vitamin A and high levels of free radicals may be the cause. This effect is thought to be specific to cats and has not been reproduced in any other animal. To produce these effects, the cats were fed solely on food that was irradiated at a dose at least five times higher than the maximum allowable dose.

It may seem reasonable to assume that irradiating food might lead to radiation-tolerant strains, similar to the way that strains of bacteria have developed resistance to antibiotics. Bacteria develop a resistance to antibiotics after an individual uses antibiotics repeatedly. Much like pasteurization plants, products that pass through irradiation plants are processed once, and are not processed and reprocessed. Cycles of heat treatment have been shown to produce heat-tolerant bacteria, yet no problems have appeared so far in pasteurization plants. Furthermore, when the irradiation dose is chosen to target a specific species of microbe, it is calibrated to doses several times the value required to target the species. This ensures that the process randomly destroys all members of a target species. Therefore, the more irradiation-tolerant members of the target species are not given any evolutionary advantage. Without evolutionary advantage, selection does not occur. As to the irradiation process directly producing mutations that lead to more virulent, radiation-resistant strains, the European Commission's Scientific Committee on Food found that there is no evidence; on the contrary, irradiation has been found to cause loss of virulence and infectivity, as mutants are usually less competitive and less adapted.

Misconceptions

Some who advocate against food irradiation argue the safety of irradiated food is not scientifically proven because there are a lack of long-term studies in spite of the fact that hundreds of animal feeding studies of irradiated food, including multigenerational studies, have been performed since 1950. Endpoints investigated have included sub chronic and chronic changes in metabolism, histopathology, function of most systems, reproductive effects, growth, teratogenicity, and mutagenicity. A large number of studies have been performed; meta-studies have supported the safety of irradiated food.

The below experiments are cited by food irradiation opponents, but either could not be verified in later experiments, could not be clearly attributed to the radiation effect, or could be attributed to an inappropriate design of the experiment:

- India's National Institute of Nutrition (NIN) found an elevated rate of cells with more than one set of genes (polyploidy) in humans and animals when fed wheat that was irradiated recently (within 12 weeks). Upon analysis, scientists determined that the techniques used by the NIN allowed for too much human error and statistical variation; therefore, the results were unreliable. After multiple studies by independent agencies and scientists, no correlation between polyploidy and irradiation of food could be found.

Indirect Effects of Irradiation

The indirect effects of irradiation are the concerns and benefits of irradiation that are related to how making food irradiation a common process will change the world, with emphasis on the system of food production.

If irradiation was to become common in the food handling process there would be a reduction of the prevalence of foodborne illness and potentially the eradication of specific pathogens. However, multiple studies suggest that an increased rate of pathogen growth may occur when irradiated food is cross-contaminated with a pathogen, as the competing spoilage organisms are no longer present. This being said, cross contamination itself becomes less prevalent with an increase in usage of irradiated foods.

The ability to remove bacterial contamination through post-processing by irradiation may reduce the fear of mishandling food which could cultivate a cavalier attitude toward hygiene and result in contaminants other than bacteria. However, concerns that the pasteurization of milk would lead to increased contamination of milk were prevalent when mandatory pasteurization was introduced, but these fears never materialized after adoption of this law. Therefore, it is unlikely for irradiation to cause an increase of illness due to non bacteria-based contamination.

Treatment

Up to the point where the food is processed by irradiation, the food is processed in the same way as all other food. To treat the food, they are exposed to a radioactive source, for a set period of time to achieve a desired dose. Radiation may be emitted by a radioactive substance, or by X-ray and electron beam accelerators. Special precautions are taken to ensure the food stuffs never come in contact with the radioactive substances and that the personnel and the environment are protected from exposure radiation.Irradiation treatments are typically classified by dose (high, medium, and low), but are sometimes classified by the effects of the treatment (radappertization, radicidation and radurization). Food irradiation is sometimes referred to as "cold pasteurization" or "electronic pasteurization" because ionizing the food does not heat the food to high temperatures during the process, and the effect is similar to heat pasteurization. The term "cold pasteurization" is controversial because the term may be used to disguise the fact the food has been irradiated and pasteurization and irradiation are fundamentally different processes.

Treatment costs vary as a function of dose and facility usage. A pallet or tote is typically exposed for several minutes to hours depending on dose. Low-dose applications such as disinfestation of fruit range between US$0.01/lbs and US$0.08/lbs while higher-dose applications can cost as much as US$0.20/lbs.

Packaging

Food processors and manufacturers today struggle with using affordable, efficient packaging materials for irradiation based processing. The implementation of irradiation on prepackaged foods has been found to impact foods by inducing specific chemical alterations to the food packaging material that migrates into the food. Cross-linking in various plastics can lead to physical and chemical modifications that can increase the overall molecular weight. On the other hand, chain scission is fragmentation of polymer chains that leads to a molecular weight reduction.

Dosimetry

The radiation absorbed dose is the amount energy absorbed per unit weight of the target material. Dose is used because, when the same substance is given the same dose, similar changes are observed in the target material. The SI unit for dose is grays (Gy or J/kg). Dosimeters are used to measure dose, and are small components that, when exposed to ionizing radiation, change measurable physical attributes to a degree that can be correlated to the dose received. Measuring dose (dosimetry) involves exposing one or more dosimeters along with the target material.

For purposes of legislation doses are divided into low (up to 1 kGy), medium (1 kGy to 10 kGy), and high-dose applications (above 10 kGy). High-dose applications are above those currently permitted in the US for commercial food items by the FDA and other regulators around the world. Though these doses are approved for non commercial applications, such as sterilizing frozen meat for NASA astronauts (doses of 44 kGy) and food for hospital patients.

Applications of food irradiation		
	Application	Dose (kGy)
Low dose (up to 1 kGy)	Inhibit sprouting (potatoes, onions, yams, garlic)	0.06 - 0.2
	Delay in ripening (strawberries, potatoes)	0.5 - 1.0
	Prevent insect infestation (grains, cereals, coffee beans, spices, dried nuts, dried fruits, dried fish, mangoes, papayas)	0.15 - 1.0
	Parasite control and inactivation (tape worm, trichina)	0.3 - 1.0
Medium dose (1 kGy to 10 kGy)	Extend shelf-life (raw and fresh fish, seafood, fresh produce, refrigerated and frozen meat products)	1.0 - 7.0
	Reduce risk of pathogenic and spoilage microbes (meat, seafood, spices, and poultry)	1.0 - 7.0
	Increased juice yield, reduction in cooking time of dried vegetables	3.0 - 7.0
High dose (above 10 kGy)	Enzymes (dehydrated)	10.0
	Sterilization of spices, dry vegetable seasonings	30.0 max
	Sterilization of packaging material	10.0 - 25.0
	Sterilization of foods (NASA and hospitals)	44.0

Processes

Gamma Irradiation

Gamma irradiation is produced from the radioisotopes cobalt-60 and caesium-137, which are derived by neutron bombardment of cobalt-59 and as a nuclear source by-product, respectively. Cobalt-60 is the most common source of gamma rays for food irradiation in commercial scale facilities as it is water insoluble and hence has little risk of environmental contamination by leakage into the water systems. As for transportation of the radiation source, cobalt-60 is transported in special trucks that prevent release of radiation and meet standards mentioned in the Regulations for Safe Transport of Radioactive Materials of the International Atomic Energy Act. The special trucks must meet high safety standards and pass extensive tests to be approved to ship radiation sources. Conversely, caesium-137, is water soluble and poses a risk of environmental contamination. Insufficient quantities are available for large scale commercial use. An incident where water-soluble caesium-137 leaked into the source storage pool requiring NRC intervention has led to near elimination of this radioisotope.

Cobalt 60 stored in Gamma Irradiation machine.

Gamma irradiation is widely used due to its high penetration depth and dose uniformity, allowing for large-scale applications with high through puts. Additionally, gamma irradiation is significantly less expensive than using an X-ray source In most designs, the radioisotope, contained in stainless steel pencils, is stored in a water-filled storage pool which absorbs the radiation energy when not in use. For treatment, the source is lifted out of the storage tank, and product contained in totes is passed around the pencils to achieve required processing.

Electron Beam

Treatment of electron beams is created as a result of high energy electrons in an accelerator that generates electrons accelerated to 99% the speed of light. This system uses electrical energy and can be powered on and off. The high power correlates with a higher throughput and lower unit cost, but electron beams have low dose uniformity and a penetration depth of centimeters. Therefore, electron beam treatment works for products that have low thickness.

Irradiated Guava: Spring Valley Fruits, Mexico.

X-ray

X-rays are produced by bombardment of dense target material with high energy accelerated electrons(this process is known as bremsstrahlung-conversion), giving rise to a continuous energy spectrum. Heavy metals, such as tantalum and tungsten, are used because of their high atomic numbers and high melting temperatures. Tantalum is usually preferred versus tungsten for industrial, large-area, high-power targets because it is more workable than tungsten and has a higher threshold energy for induced reactions. Like electron beams, x-rays do not require the use of radioactive materials and can be turned off when not in use. X-rays have high penetration depths and high dose uniformity but they are a very expensive source of irradiation as only 8% of the incident energy is converted into X-rays.

Cost

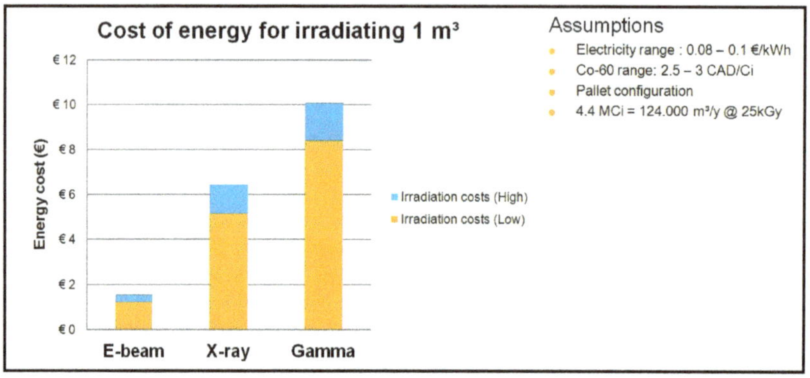

Efficiency illustration of the different radiation technologies (electron beam, X-ray, gamma rays).

The cost of food irradiation is influenced by dose requirements, the food's tolerance of radiation, handling conditions, i.e., packaging and stacking requirements, construction costs, financing arrangements, and other variables particular to the situation. Irradiation is a capital-intensive technology requiring a substantial initial investment, ranging from $1 million to $5 million. In the case of large research or contract irradiation facilities, major capital costs include a radiation source, hardware (irradiator, totes and conveyors, control systems, and other auxiliary equipment), land (1 to 1.5 acres), radiation shield, and warehouse. Operating costs include salaries (for fixed and variable labor), utilities, maintenance, taxes/insurance, cobalt-60 replenishment, general utilities, and miscellaneous operating costs. Perishable food items, like fruits, vegetables and meats would still require to be handled in the cold chain, so all other supply chain costs remain the same.

Public Perception

Negative connotations associated with the word "radiation" are thought to be responsible for low consumer acceptance. Several national expert groups and two international expert groups evaluated the available data and concluded that any food at any dose is wholesome and safe to consume.

Irradiation has been approved by many countries. For example, in the U.S. the FDA has approved food irradiation for over fifty years. However, in the past decade the major growth area is for fruits and vegetables that are irradiated to prevent the spread of pests. In the early 2000s in the US, irradiated meat was common at some grocery stores, but because of lack of consumer demand, it is no longer common. Because consumer demand for irradiated food is low, reducing the spoilage between manufacturer and consumer purchase and reducing the risk of food borne illness is currently not sufficient incentive for most manufacturers to supplement their process with irradiation. Nevertheless, food irradiation does take place commercially and volumes are in general increasing at a slow rate, even in the European Union where all member countries allow the irradiation of dried herbs spices and vegetable seasonings but only a few allow other foods to be sold as irradiated.

Although there are some consumers who choose not to purchase irradiated food, a sufficient market has existed for retailers to have continuously stocked irradiated products for years. When labeled irradiated food is offered for retail sale, these consumers buy it and re-purchase it, indicating that it is possible to successfully market irradiated foods, therefore retailers not stocking irradiated foods might be a major bottleneck to the wider adoption of irradiated foods. It is however, widely believed that consumer perception of foods treated with irradiation is more negative than those processed by other means and some industry studies indicate the number of consumers concerned about the safety of irradiated food decreased between 1985 and 1995 to levels comparable to those of people concerned about food additives and preservatives. Even though is it is untrue "People think the product is radioactive," said Harlan Clemmons, president of Sadex, a food irradiation company based in Sioux City, Iowa. Because of these concerns and the increased cost of irradiated foods, there is not a widespread public demand for the irradiation of foods for human consumption. Irradiated food does not become radioactive.

Standards and Regulations

The Codex Alimentarius represents the global standard for irradiation of food, in particular under the WTO-agreement. Regardless of treatment source, all processing facilities must adhere to safety standards set by the International Atomic Energy Agency (IAEA), Codex Code of Practice for the Radiation Processing of Food, Nuclear Regulatory Commission (NRC), and the International Organization for Standardization (ISO). More specifically, ISO 14470 and ISO 9001 provide in-depth information regarding safety in irradiation facilities.

All commercial irradiation facilities contain safety systems are designed to prevent exposure of personnel to radiation. The radiation source is constantly shielded by water, concrete, or metal. Irradiation facilities are designed with overlapping layers of protection, interlocks, and safeguards to prevent accidental radiation exposure. Additionally, "melt-downs" do not occur in facilities because the radiation source gives off radiation and decay heat; however, the heat is not sufficient to melt any material.

Labeling

The provisions of the Codex Alimentarius are that any "first generation" product must be labeled "irradiated" as any product derived directly from an irradiated raw material; for ingredients the provision is that even the last molecule of an irradiated ingredient must be listed with the ingredients even in cases where the unirradiated ingredient does not appear on the label. The RADURA-logo is optional; several countries use a graphical version that differs from the Codex-version. The suggested rules for labeling is published at CODEX-STAN – 1 (2005), and includes the usage of the Radura symbol for all products that contain irradiated foods. The Radura symbol is not a designator of quality. The amount of pathogens remaining is based upon dose and the original content and the dose applied can vary on a product by product basis.

The European Union follows the Codex's provision to label irradiated ingredients down to the last molecule of irradiated food. The European Community does not provide for the use of the Radura logo and relies exclusively on labeling by the appropriate phrases in the respective languages of the Member States. The European Union enforces its irradiation labeling laws by requiring its member countries to perform tests on a cross section of food items in the market-place and to report to the European Commission. The results are published annually in the OJ of the European Communities.

The Radura symbol, as required by U.S. Food and Drug Administration regulations to show a food has been treated with ionizing radiation.

The US defines irradiated foods as foods in which the irradiation causes a material change in the food, or a material change in the consequences that may result from the use of the food. Therefore, food that is processed as an ingredient by a restaurant or food processor is exempt from the labeling requirement in the US. All irradiated foods must include a prominent Radura symbol followed in addition to the statement "treated with irradiation" or "treated by irradiation. Bulk foods must be individually labeled with the symbol and statement or, alternatively, the Radura and statement should be located next to the sale container.

Packaging

Under section 409 of the Federal Food, Drug, and Cosmetic Act, irradiation of prepackaged foods requires premarket approval for not only the irradiation source for a specific food but also for the food packaging material. Approved packaging materials include various plastic films, yet does not cover a variety of polymers and adhesive based materials that have been found to meet specific standards. The lack of packaging material approval limits manufacturers production and expansion of irradiated prepackaged foods.

Approved materials by FDA for Irradiation according to 21 CFR 179.45:

Material	Paper (kraft)	Paper (glassine)	Paper-board	Cello-phane (coated)	Polyolefin film	Polyestyrene film	Ny-lon-6	Vegetable Parchment	Nylon 11
Irradiation (kGy)	.05	10	10	10	10	10	10	60	60

Food Safety

In 2003, the Codex Alimentarius removed any upper dose limit for food irradiation as well as clearances for specific foods, declaring that all are safe to irradiate. Countries such as Pakistan and Brazil have adopted the Codex without any reservation or restriction. Other countries, including New Zealand, Australia, Thailand, India, and Mexico, have permitted the irradiation of fresh fruits for fruit fly quarantine purposes, amongst others.

Standards that describe calibration and operation for radiation dosimetry, as well as procedures to relate the measured dose to the effects achieved and to report and document such results, are maintained by the American Society for Testing and Materials (ASTM international) and are also available as ISO/ASTM standards.

All of the rules involved in processing food are applied to all foods before they are irradiated.

The U.S. Food and Drug Administration (FDA) is the agency responsible for regulation of radiation sources in the United States. Irradiation, as defined by the FDA is a "food additive" as opposed to a food process and therefore falls under the food additive regulations. Each food approved for irradiation has specific guidelines in terms of minimum and maximum dosage as deterred safe by the FDA. Packaging materials containing the food processed by irradiation must also undergo approval. The United States Department of Agriculture (USDA) amends these rules for use with meat, poultry, and fresh fruit.

The United States Department of Agriculture (USDA) has approved the use of low-level irradiation as an alternative treatment to pesticides for fruits and vegetables that are considered hosts to a number of insect pests, including fruit flies and seed weevils. Under bilateral agreements that allows less-developed countries to earn income through food exports agreements are made to allow them to irradiate fruits and vegetables at low doses to kill insects, so that the food can avoid quarantine.

European law dictates that all member countries must allow the sale of irradiated dried aromatic herbs, spices and vegetable seasonings. However, these Directives allow Member States to maintain previous clearances food categories the EC's Scientific Committee on Food (SCF) had previously approved (the approval body is now the European Food Safety Authority). Presently, Belgium, Czech Republic, France, Italy, Netherlands, Poland, and the United Kingdom allow the sale of many different types of irradiated foods. Before individual items in an approved class can be added to the approved list, studies into the toxicology of each of such food and for each of the proposed dose ranges are requested. It also states that irradiation shall not be used "as a substitute for hygiene or health practices or good manufacturing or agricultural practice". These Directives only control food irradiation for food retail and their conditions and controls are not applicable to the irradiation of food for patients requiring sterile diets.

Because of the Single Market of the EC any food, even if irradiated, must be allowed to be marketed in any other Member State even if a general ban of food irradiation prevails, under the condition that the food has been irradiated legally in the state of origin. Furthermore, imports into the EC are possible from third countries if the irradiation facility had been inspected and approved by the EC and the treatment is legal within the EC or some Member state.

Australia banned irradiated cat food after a national scare where cats suffered from paralyzation after eating a specific brand of highly irradiated catfood for an extended period of time. The suspected culprit was malnutrition from consuming food depleted of Vitamin A by the irradiation process. The incident was linked only to a single batch of one brand's product and no illness was linked to any of that brand's other irradiated batches of the same product or to any other brand of irradiated cat food. This, along with incomplete evidence indicating that the cat food was not sufficiently depleted of Vitamin A makes irradiation a less likely cause. Further research has been able to experimentally induce the paralyzation of cats by via Vitamin A deficiency by feeding highly irradiated food. For more details see the Long term impacts section.

Nuclear Safety and Security

Interlocks and safeguards are mandated to minimize this risk. There have been radiation-related accidents, deaths, and injury at such facilities, many of them caused by operators overriding the safety related interlocks. In a radiation processing facility, radiation specific concerns are supervised by special authorities, while "Ordinary" occupational safety regulations are handled much like other businesses.

The safety of irradiation facilities is regulated by the United Nations International Atomic Energy Agency and monitored by the different national Nuclear Regulatory Commissions. The regulators enforce a safety culture that mandates that all incidents that occur are documented and thoroughly analyzed to determine the cause and improvement potential. Such incidents are studied by personnel at multiple facilities, and improvements are mandated to retrofit existing facilities and future design.

In the US the Nuclear Regulatory Commission (NRC) regulates the safety of the processing facility, and the United States Department of Transportation (DOT) regulates the safe transport of the radioactive sources.

Irradiated Food Supply

As of 2010, the quantities of foods irradiated in Asia, the EU and the US were 285,200, 9,300, and 103,000 tons. Authorities in some countries use tests that can detect the irradiation of food items to enforce labeling standards and to bolster consumer confidence. The European Union monitors the market to determine the quantity of irradiated foods, if irradiated foods are labeled as irradiated, and if the irradiation is performed at approved facilities.

Irradiation of fruits and vegetables to prevent the spread of pest and diseases across borders has been increasing globally. In 2010, 18,446 tonnes of fruits and vegetables were irradiated in six countries for export quarantine control. 97% of this was exported to the United States.

In total, 103 000 tonnes of food products were irradiated on mainland United States in 2010. The three types of foods irradiated the most were spices (77.7%), fruits and vegetables (14.6%) and meat and poultry (7.77%). 17 953 tonnes of irradiated fruits and vegetables were exported to the mainland United States. Mexico, the United States' state of Hawaii, Thailand, Vietnam and India export irradiated produce to the mainland U.S. Mexico, followed by the United States' state of Hawaii, is the largest exporter of irradiated produce to the mainland U.S.

In total, 6 876 tonnes of food products were irradiated in European Union countries in 2013; mainly in four member state countries: Belgium (49.4%), the Netherlands (24.4%), Spain (12.7%)

and France (10.0%). The two types of foods irradiated the most were frog legs (46%), and dried herbs and spices (25%). There has been a decrease of 14% in the total quantity of products irradiated in the EU compared to the previous year 2012 (7 972 tonnes).

Pulsed Electric Field

Pulsed electric field (PEF) technology is a non-thermal food preservation method that involves the use of short electricity pulses for microbial inactivation while imposing minimal detrimental influence on food quality. This technology has the major advantage to provide high-quality foods to the consumers. PEF is claimed as superior to thermal processing and preservation methods because it reduces detrimental changes in food quality and nutrition and keeps physical and sensorial attributes of food.

PEF technology has wide range of applications ranging from liquid or semi-solid foods to solid foods. Most PEF studies have captivated on application of high voltage pulses for microbial inactivation milk and dairy products, egg products, juices and other liquid foods . However, most researchers have mainly focused on the aspect of food preservation with special reference to the microbial control and a lesser information is presented about the effect of PEF on food composition, quality and acceptability. Recently, some investigations have been conducted to evaluate the potential of PEF for improving food processing efficiency like enhancement of juice extraction and escalation of the food dehydration or drying.

Wouters et al. reported that PEF is known as one of the most auspicious non-thermal tools for microbial decontamination of foods. It involves the generation of electric fields(5-50 kV/cm) with the help of short high voltage pulses (us) between two electrodes that leads to microbial inactivation at temperatures lower than thermal methods. They also proposed that exact mechanistic approach of microbial inactivation is not known however, it is normally hypothesized that PEF causes penneabilization or depolarization of microbial membranes.

Studies revealed that PEF technology enables inactivation of bacterial and yeast vegetative cells in various foods. On the contrary, bacterial spores cannot be killed by employing pulsed electric fields because spores are resistant to PEF. Thus, applications of PEF are primarily focused on pathogenic and spoilage causing microorganisms in food. In addition to the pronounced effect of this technology in controlling microbiological spoilage of foods in a rapid and uniform manner, PEF also delivers shelf life extension without using heat treatment and preservation of sensorial and nutritional quality of foods. Likewise, PEF is also capable to improve the energy usage in an efficient and economical way. Hence, successful applications of PEF technology propose an alternative to conventional thermal processing techniques for food preservation and processing. In this topic, principles, mechanisms and recent applications of PEF technology are reviewed and compared with thermal processing technologies. Furthermore, examples are coded to illustrate the potential of PEF technology for aiming at preserving the quality features of various foods.

Principle of PEF

PEF technology involves the use of pulses having higher electric fields for only a few micro to milliseconds with intensity in the range of 10-80 kV/cm. The process depends on the number of pulses delivered to the product which is held between two electrodes. These electrodes have a specific gap between them which is known as treatment gap of the chamber. During PEF processing, high voltage is applied that results in the inactivation of microorganisms present in the food sample. The electric field is applied in different forms like as exponentially decaying waves, bipolar waves or oscillatory pulses. The process can also be carried at various temperature ranges such as ambient, sub-ambient and above-ambient temperatures. Food is packed after treatment with PEF and then stored under refrigerated conditions.

The science involved behind the transfer of electric pulses from food is that food contains several ions that provide a definite level of electrical conductivity to the product. This technique is usually preferred for liquid foods because electrical current flows into the liquid food more efficiently and the transfer of pulses from one point to other in liquids is quite easy due to the presence of charged molecules present.

A group of researchers, Zimmermann & Benz stated that mechanism of the functioning of PEF technology is the delivery of pulsing power to the product that is placed between a set of electrodes confining the treatment gap of the PEF chamber. The typical system has a pulse generator that produces high voltage pulses, treatment chamber that handles the product to be treated and associated with controlling and monitoring devices. Food product is placed in the treatment chamber equipped with electrodes connected with each other by a nonconductive material which prevents the flow of electricity from one to the other. High voltage electrical pulses are generated that are transferred to the product. The product placed between the electrodes experiences a force per unit charge which ruptures the bacterial cell membranes . In general, PEF technology is suggested for pasteurization of various food products including milk, juices, yogurt, liquid eggs and soups . Furthermore, combination of PEF with ultrasound, high pressure and ultraviolet light treatments may enhance the process output.

Microbial Inactivation by PEF

Several studies have investigated the mode of action of PEF to reduce the microbial load in various food products. Nonetheless, the exact mechanistic approach underlying the microbial inactivation by PEF has not been fully expounded as yet. However, a general mechanism of PEF action involves instability of microbial membranes by the induction of electrical field and electromechanical compression that leads to the pore formation in membrane . Mechanical fickleness of membranes is caused due to critical membrane potential which is formed by electrical field. Electroporation results in a significant increase in the membrane rupture and permeability which is termed as electro permeabilization. Electro permeabilization can be reversible or irreversible, depending on the degree of membrane organizational changes that results in cell death. Literature explains that membrane permeability is increased in a considerable manner by increasing the strength of electric field. This elevated membrane instability harmonizes with inactivation of microbial cells. In general, spores are claimed as more obstinate to PEF treatment as compared to vegetative cells.

Factors Influencing PEF Performance to Inactivate Microorganisms

The capability of PEF to inactivate microbes depends on several factors that can be categorized as process parameters, nature of product and properties of microbial cells . These factors play a dynamic role to attain the optimum performance of PEF treatment. An overview of these parameters is discussed herein.

Process Parameters

Several process parameters affect the ability of PEF to reduce microbial population in food such as strength of electric field, pulse length and shape, number of pulses and temperature. On a general node, increased intensity of these factors improves microbial inactivation but their exact relationship with the survival rate of microorganisms is not clear. Therefore, exact measurement of all these parameters is required to acquire reliable results.

Product Nature

The administration of PEF is also influenced by nature of product as a vast range of products are being treated by PEF which include fruit juices, liquid eggs, milk and dry herbs. It is investigated through experimental trials that PEF treatment is not much effective for products having particles or special structures, i.e. emulsions, Additionally, physical and chemical properties of the food also affect the rate of microbial decontamination. Various studies have revealed the influence of pH, water activity and electrical conductivity on the efficiency of PEF to inactivate microorganisms.

pH has a significant influence on the inactivation kinetics of microbes. Jeantet et al. reported that higher inactivation of Salmonella was observed in foods having neutral or above neutral pH values. Similarly, Alvarez et al. testified substantial reduction in L. mono cytogenes number in high acid foods such as citrus juices. Likewise, conductivity of the treatment medium has inverse relation with microbial inactivation. It is observed that foods having high electrical conductivity show less inactivation of microorganisms after PEF treatment. On the contrary, water activity has a direct relation with microbial reduction by PEF treatment as confirmed by Min & Zhang.

Characteristics of Microbes

Inactivation of microorganism by employing PEF technology also depends on microbial characteristics including type of microorganisms, species and strains. Generally, Gram- positive and Gram-negative bacteria are thought to be more resistant in comparison with yeast cells. Likewise, bacterial and mold spores are also claimed as recalcitrant to PEF treatment. Additionally, cell size and shape also affect the inactivation kinetics due to the difference in the development of critical membrane potential. PEF treatment affects different bacterial species at altered rate. It is usually proposed that Salmonella and E. coli are more susceptible to PEF as compared to Listeria and Bacillus species. Growth conditions like temperature, growth medium, concentration of nutrients and pH of the treated medium also influence PEF efficiency as well.

Applications in Food Industry

Application of PEF technology has been extensively demonstrated for the pasteurization of various food products like juices, milk & dairy products, soup and liquid eggs. However, it has several limitations such as product must be free from air bubbles and must have lower electrical conductivity. Additionally, particle size should be less than the gap of the treatment region to ensure appropriate treatment. PEF is generally not suitable for solid foods however several solid products have also been investigated to be efficiently treated by deploying PEF treatment. PEF technology can also be used to enhance the extraction of several bioactive components and sugars from plant cells.

PEF processing has shown its potential to treat less viscous fruit juices having less electrical conductivity such as apple, citrus and cranberry juices. PEF technology also executes beneficial aspects on the quality parameters of fruit juice. Correspondingly, Yeom et al. compared pasteurized and PEF-treated citrus juice during refrigerated storage (4°C) for a period of 112 days and observed less browning in PEF-processed juice comparatively to traditionally pasteurized juice due to conversion of ascorbic acid to furfural. Besides, PEF-treated foods also retained their fresh flavor, textural & functional attributes and extended shelf life in addition with microbiological safety . In recent years, PEF technology has been utilized for various purposes like enhancement of drying efficiency, modification of enzymatic activity, solid food preservation, waste water treatment and extraction.

PEF in Fruit Processing

PEF processing has promising applications in citrus industry with special reference to the inactivation of microorganisms and prevention of developing off-flavors during the storage . Jemai & Vorobiev noticed that treatment of apple juice with PEF resulted in enhancement of diffusion coefficients of soluble substances. This technology can also be fruitfully applied for the disintegration of biological tissues that enhances the extraction of intracellular compounds from different fruits. For instance, pectin is a very useful component found in many fruits that is traditionally being extracted through enzymatic reaction but this reaction has less yield of pectin due to poor efficiency. Alternatively, PEF treatment is employed as short pulses to avoid excessive heat and undesirable electrolytic reactions that can enhance the extraction rate of pectin from fruit pomace.

Bacterial Inactivation in Milk by PEF

Inadequately pasteurized milk may cause several health hazards due to the presence of several spoilage causing and pathogenic bacteria particularly Escherichia coli and Listeria and Pseudomonas spp . Elevated concerns regarding the impact of heat treatment on the quality of milk and consumer demand for fresh-like quality attribute products have encouraged the induction of PEF for milk pasteurization. PEF-induced microbial inactivation is believed to be an effective way of milk preservation without adversely affecting the milk quality.

Studies have demonstrated the effectiveness of PEF processing for microbial reduction in simulated milk ultra-filtrate and skim milk . However, presence of fat and protein moieties limits the adeptness of PEF in whole milk because these molecules protect bacterial cells during treatment . Therefore, it is imperative to validate the worth of PEF treatment to inactivate bacteria in a complex whole milk matrix for a genuine comparison with thermal pasteurization.

Garcia et al. observed sub-lethal injury in bacterial cells and concluded that PEF treatment can be successfully used in synergy with other hurdles to get more benefits. In another trial, PEF processing of milk was combined with heat treatment up to 55-60°C and a significant reduction was observed in the microbial load . More recently, Sharma et al. employed PEF treatment at different dose, time and temperature ranges to inactivate Gram-positive and Gram-negative bacteria in whole milk and reported 5-6 log reduction in bacterial number at 22-28 kV cm-1 for 17-101 μs at 50°C. On the other hand, some studies did not clearly demonstrate the effects pre-heating, temperature, time & dose variation on PEF efficiency and effects of PEF treatment on milk quality and functional properties. Therefore, it is questionable whether PEF can maintain the integrity of heat-sensitive milk components while inactivating spoilage and pathogenic microorganisms in whole milk and need further exploration.

PEF and Meat Quality

Meat products have been widely consumed around the globe due to the presence of valuable micronutrients high quality protein . Quality of meat is at prime importance because meat quality is considered as the most vital factor for purchasing decisions of consumers . Pulsed electric field technology has shown its potential for various applications in solid foods with different aims including structural modifications, changing physical quality parameters, extraction of bioactive compounds and preservation. Nevertheless, this technology has limited applications in muscle foods . PEF technology can be employed in meat processing for various purposes including enhancement of cell permeation to increase tenderness, attenuation of microbial load to improve the shelf life and maintenance of volatile profile of meat during storage.

PEF technology can considerably improve enzyme release and glycolysis that are essential for proteolysis for meat tenderization. O'Dowd et al. studied the effect of PEF on meat quality @ 1.1-2.8 kV cm−1 and reported no significant improvement in meat tenderness during storage. Later on, Bekhit et al. found a weighty improvement in beef tenderness by PEF treatment (5-10 kV). Likewise, applied PEF (5-10 kV) at different frequencies (20, 50 and 90 Hz) on beef muscles and revealed that PEF treatment reduced shear force up to 19% & improved tenderness by augmenting the degradation of desmin and troponin-T during a refrigeration storage of 21 days.

Modified Atmosphere

Modified Atmosphere Packaging is also known as gas flushing, protective atmosphere packaging or reduced oxygen packaging.

Modified Atmosphere enables fresh and minimally processed packaged food products to maintain visual, textural and nutritional appeal. The controlled MAP environment enables food packaging to provide an extended shelf life without requiring the addition of chemical preservatives or stabilisers. Processors and marketers of food products rely on Modified Atmosphere Packaging to assure fresh and flavourful products that continually meet the consumer's expectation for brand quality, consistency, freshness and in-stock availability.

Modified Atmosphere Packaging is an optimal blend of pure oxygen, carbon dioxide and nitrogen within a high barrier or permeable package. A finely adjusted and carefully controlled gas blend is developed to meet the specific respiration needs for each packaged food product.

Plastic films, foils and other packaging materials that demonstrate specified gas permeability properties and/or water vapour permeability properties are selected for use. These high barrier substrates become MAP Packages after they are formed into trays, lid stock or bags and filled with a select blend of oxygen, carbon dioxide and nitrogen environmental gasses.

Packaging films are selected to match the characteristics and needs of the food product. Film permeability, water vapour transmission rates and sealing characteristics need to be measured and tested at film selection and again at package converting and product fill stages, since the ability of a film to handle MAP performance characteristics may vary within each stage.

Working of Modified Atmosphere Packaging

The Modified Atmosphere Package environment is formed from a finely balanced mix of normal atmospheric gases. The finely balanced MAP gas mix slows down the product aging process to reduce colour loss, odour and off-taste resulting from product deterioration, spoilage and rancidity caused by mold and other anaerobic organisms.

A carefully controlled Modified Atmosphere Package achieves and maintains an optimal respiration rate to preserve the fresh colour, taste and nutrient content of red meat, seafood, minimally processed fruits and vegetables, pasta, prepared foods, cheese, baked goods, cured meats and dried foods throughout an extended shelf life.

Modified Atmosphere Packaging Offers Supply Chain Efficiencies

Longer shelf life MAP packages allow food processors, food manufacturers, food distributors and food retailers to better control product quality, availability and costs.

Longer freshness cycles permit grocers to eliminate frequent product rotation, removal and restocking; thereby reducing labour and waste disposal costs.

Distributors can extend distribution territories or offer a greater variety of product lines to the retailer, since less frequent product replacement requirements permits growth in other areas.

Food manufacturers are able to take advantage of extended replacement cycles to reduce production replacement demands. Manufacturing capacity can be more profitably utilised by developing and offering new products.

The first recorded beneficial effects of using modified atmosphere date back to 1821. Jacques Etienne Berard, a professor at the School of Pharmacy in Montpellier, France, reported delayed ripening of fruit and increased shelf life in low-oxygen storage conditions. Controlled Atmosphere Storage (CAS) was used from the 1930's when ships transporting fresh apples and pears had high levels of CO_2 in their holding rooms in order to increase the shelf life of the product. In the 1970s MA packages reached the stores when bacon and fish were sold in retail packs in Mexico. Since then development has been continuous and interest in MAP has grown due to consumer demand.

Theory

Atmosphere within the package can be modified passively or actively. In passive modified atmosphere packaging (MAP), the high concentration of CO_2 and low O_2 levels in the package is achieved over time as a result of respiration of the product and gas transmission rates of the packaging film. This method is commonly used for fresh respiring fruits and vegetables. Reducing O_2 and increasing CO_2 slows down respiration rate, conserves stored energy, and therefore extended shelf life. On the other hand, active MA involves the use of active systems such as O_2 and CO_2 scavengers or emitters, moisture absorbers, ethylene scavengers, ethanol emitters and gas flushing in the packaging film or container to modify the atmosphere within the package.

The mixture of gases selected for a MAP package depends on the type of product, the packaging materials and the storage temperature. The atmosphere in an MA package consists mainly of adjusted amounts of N_2, O_2, and $CO_{2.}$ Reduction of O_2 promotes delay in deteriorative reactions in foods such as lipid oxidation, browning reactions and growth of spoilage organisms. Low O_2 levels of 3-5% are used to slow down respiration rate in fruits and vegetables. In the case of red meat, however, high levels of O_2 (~80%) are used to reduce oxidation of myoglobin and maintain an attractive bright red color of the meat. Meat color enhancement is not required for pork, poultry and cooked meats; therefore, a higher concentration of CO_2 is used to extend the shelf life. Levels higher than 10% of CO_2 are phytotoxic for fruit and vegetables, so CO_2 is maintained below this level. N_2 is mostly used as a filler gas to prevent pack collapse. In addition, it is also used to prevent oxidative rancidity in packaged products such as snack foods by displacing atmospheric air, especially oxygen, therefore extending shelf life. The use of noble gases such as Helium (He), Argon (Ar) and Xenon (Xe) to replace N_2 as the balancing gas in MAP can also be used to preserve and extend the shelf life of fresh and minimally processed fruits and vegetables. Their beneficial effects are due to their higher solubility and diffusivity in water, making them more effective in displacing O_2 from cellular sites and enzymatic O_2 receptors.

There has been a debate regarding the use of carbon monoxide (CO) in the packaging of red meat due to its possible toxic effect on packaging workers. Its use results in a more stable red color of carboxymyoglobin in meat, which leads to another concern that it can mask evidence of spoilage in the product.

Effect on Microorganisms

Low O_2 and high CO_2 concentrations in packages are effective in limiting the growth of Gram negative bacteria, molds and aerobic microorganisms, such as *Pseudomonas* spp. High O_2 combined with high CO_2 could have bacteriostatic and bactericidal effects by suppression of aerobes by high CO_2 and anaerobes by high O_2. CO_2 has the ability to penetrate bacterial membrane and affect intracellular pH. Therefore, lag phase and generation time of spoilage microorganisms are increased resulting in shelf life extension of refrigerated foods. Since the growth of spoilage microorganisms are suppressed by MAP, the ability of the pathogens to grow is potentially increased. Microorganisms that can survive under low oxygen environment such as *Campylobacter jejuni, Clostridium botulinum, E. coli, Salmonella, Listeria* and *Aeromonas hydophila* are of major concern for MA packaged products. Products may appear organoleptically acceptable due to the delayed growth of the spoilage microorganisms but might contain harmful pathogens. This risk can be minimized by use of additional hurdles such as temperature control (maintain temperature below 3 degrees C), lowering water activity (less than 0.92), reducing pH (below 4.5) or addition or preservatives such as nitrite to delay metabolic activity and growth of pathogens.

Packaging Materials

Flexible films are commonly used for products such as fresh produce, meats, fish and bread seeing as they provide suitable permeability for gases and water vapor to reach the desired atmosphere. Pre-formed trays are formed and sent to the food packaging facility where they are filled. The package headspace then undergoes modification and sealing. Pre-formed trays are usually more flexible and allow for a broader range of sizes as opposed to thermoformed packaging materials as different tray sizes and colors can be handled without the risk of damaging the package. Thermoformed packaging however is received in the food packaging facility as a roll of sheets. Each sheet is subjected to heat and pressure, and is formed at the packaging station. Following the forming, the package is filled with the product, and then sealed. The advantages that thermoformed packaging materials have over pre-formed trays are mainly cost-related: thermoformed packaging uses 30% to 50% less material, and they are transported as rolls of material. This will amount in significant reduction of manufacturing and transportation costs.

When selecting packaging films for MAP of fruits and vegetables the main characteristics to consider are gas permeability, water vapour transmission rate, mechanical properties, transparency, type of package and sealing reliability. Traditionally used packaging films like LDPE (low-density polyethylene), PVC (polyvinyl chloride), EVA (ethylene-vinyl acetate) and OPP (oriented polypropylene) are not permeable enough for highly respiring products like fresh-cut produce, mushrooms and broccoli. As fruits and vegetables are respiring products, there is a need to transmit gases through the film. Films designed with these properties are called permeable films. Other films, called barrier films, are designed to prevent the exchange of gases and are mainly used with non-respiring products like meat and fish.

MAP films developed to control the humidity level as well as the gas composition in the sealed package are beneficial for the prolonged storage of fresh fruits, vegetables and herbs that are sensitive to moisture. These films are commonly referred to as modified atmosphere/modified humidity packaging (MA/MH) films.

Equipment

In using form-fill-seal packaging machines, the main function is to place the product in a flexible pouch suitable for the desired characteristics of the final product. These pouches can either be preformed or thermoformed. The food is introduced into the pouch, the composition of the headspace atmosphere is changed within the package; it is then heat sealed. These types of machines are typically called pillow-wrap, which horizontally or vertically form, fill and seal the product. Form-fill-seal packaging machines are usually used for large scale operations.

In contrast, chamber machines are used for batch processes. A filled pre-formed wrap is filled with the product and introduced into a cavity. The cavity is closed and vacuum is then pulled on the chamber and the modified atmosphere is inserted as desired. Sealing of the package is done through heated sealing bars, and the product is then removed. This batch process is labor intensive and thus requires a longer period of time; however, it is relatively cheaper than packaging machines which are automated.

Additionally, snorkel machines are used to modify the atmosphere within a package after the food has been filled. The product is placed in the packaging material and positioned into the machine without the need of a chamber. A nozzle, which is the snorkel, is then inserted into the packaging material. It pulls a vacuum and then flushes the modified atmosphere into the package. The nozzle is removed and the package is heat sealed. This method is suitable for bulk and large operations.

Products

Many products such as red meat, seafood, minimally processed fruits and vegetables, salads, pasta, cheese, bakery goods, poultry, cooked and cured meats, ready meals and dried foods are packaged under MA. A summary of optimal gas mixtures for MA products are shown in table.

Modified Atmosphere Packaging for different food products and optimal gas mixtures.

Product	Oxygen (%)	Carbon Dioxide (%)	Nitrogen (%)
Red Meat	80 - 85	15	-
Poultry	25	75	-
Fish	-	60	40
Soft Cheeses	-	100	-
Hard Cheeses	-	100	-
Bread	70	30	-
Fresh Pasta	-	-	100
Fruits and Vegetables	3 - 5	3 - 5	85 – 95

Nonthermal Plasma

Processed foods such as nutritional losses and adverse effects on organoleptic quality. This has led to the emergence of so called “Nonthermal Technologies”. Nonthermal technologies are preservation treatments that are effective at ambient or sub-lethal temperatures, thereby minimizing negative thermal effects on nutritional and quality parameters of food. These include the application of gamma irradiation, beta irradiation (electron beam), power ultrasound, ozonation, pulsed light, UV treatment, pulsed electric field (PEF), high hydrostatic pressure etc. Nevertheless, even some of these commercially viable inactivation methods are limited in practice due to adverse perceptions associated (like with treatments of gamma irradiation and high energy electron beams) or high initial investments required and/or other constraints. Purely physical techniques, such as high hydrostatic pressure, are chemically safe but require complex or expensive equipment, affect the quality of the product (Kruk et al. 2009) and are generally incompatible with online treatments. Very few approaches are suitable for treatment of solid foods, in particular fruit and vegetables. Very recently NTP has been added to the existing list of non-thermal processes for the decontamination of fresh produce.

Technologies like UV treatment, zonation, power ultrasound, pulsed light, electric discharge and non thermal plasma are commonly designated as Advanced Oxidation Processes (AOP). The use of pulsed UV light as a means of microbial inactivation is a mature technology that has commercial application in surface disinfection of packaging materials, but demonstrates limitations due to shadowing effects in food products. Indeed, there is currently no perfect method to achieve sterilization at ambient temperature.

High-pressure Food Preservation

High pressure processing is a promising "non-thermal" technique for food preservation that efficiently inactivates the vegetative microorganisms, most commonly related to food- borne diseases. High pressure processing is carried out with intense pressure in the range of 100-1000 MPa, with or without heat, allowing most foods to be preserved with minimal effect on taste, texture or nutritional characteristics. The main advantage of high pressure processing compared to thermal sterilization and pasteurization is maintenance of sensory and nutritional characteristic of treated food products.

Pressure treatment can be used to process both liquid and high-moisture-content solid foods. Pressure processing is a lethal to microorganisms but at relatively low temperatures (0-40°c) covalent bonds are almost unaffected. The limited effect of hPP (at moderate temperature) on covalent bonds represents a unique characteristic of this technology because hPP has a minimal effect on food chemistry. hPP provides a means for retaining food quality while avoiding the need for excessive thermal treatments or chemical preservation.

Microbial inactivation is one of the main goals for the application of high pressure technology. the inactivation effect of high pressure processing results in extending shelf- life and improving the microbial safety of food products. According to some researches high pressure treatment could be accepted as a food safety intervention for eliminating Listeria monocytogenes in processed meat products and cheese. Hydrostatic pressure treatment is also effective in inactivating other hazardous microorganisms such as E. coli, Salmonella, and Vibrio, as well as many yeasts, molds, and bacteria responsible for food spoilage. The microbiological shelf-life and food quality can be substantially extended by the use of HPP.

Bacterial spores represent a challenge for high pressure technology and more information about their resistance is required. Microbial spores suspended in foods or laboratory model system could be inactivated by high pressure treatment but compared to vegetative cells the treatment conditions must be extreme: higher pressure and long exposure time at elevated temperature. When pressure-temperature is combined at 690 MPa and 80°c for 20 minutes, the treatments was effective with a significant reduction in the Clostridium sporogenes spore count. Successful treatment of Bacillus stearothermophilus is observed were pressure treatments is combined with moderate temperature (70°C). There are no published reports on the high-pressure resistance of Clostridium botulinum spores, and their ability to withstand high pressure at low or high temperatures is unknown.

Pressure treatment can also be used to alter the functional and sensory properties of various food components, especially proteins. the tertiary and quaternary structures of molecules which are maintained mainly by hydrophobic and ionic interactions are beneficially altered by high pressure above 200 MPa. the hydrophobic and electrostatic interactions are most affected but not the hydrogen bonds which stabilize α-helical and β-pleated sheets. Meat, fish, egg and dairy proteins can be denatured with hPP in the absence of elevated temperatures. increased viscosity and opacity are obtained with little change in fresh flavour. On the other hand, high pressure has very little effect on low-molecular-weight compounds such as flavour compounds, vitamins, and pigments compared to thermal processes. Accordingly, the quality of hPP pasteurized food is very similar to that of fresh food products. the quality throughout shelf-life is influenced more by subsequent distribution and storage temperatures and the barrier properties of the packaging rather than by the high pressure treatments.

Basic High Pressure Processing Principles

High-Pressure technology has been cited as one of the best innovations in food processing from the last 50 years. Some physical and chemical changes result from application of pressure. Physical compression during pressure treatment results in a volume reduction and an increase in temperature and energy.

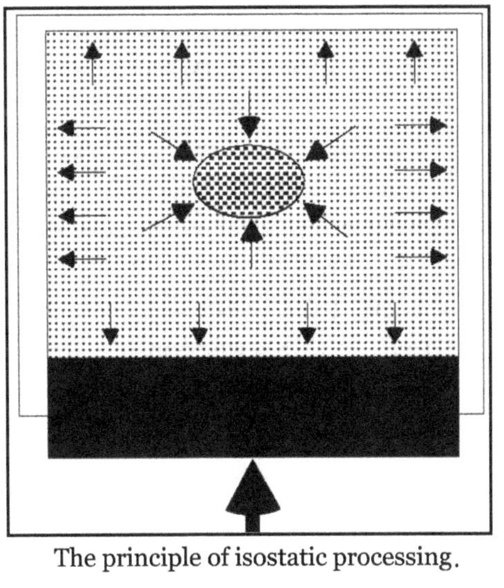

The principle of isostatic processing.

The basic principles that determine the behavior of foods under pressure are:

- Le Chatelier's principle: any reaction, conformational change, phase transition, accompanied by a decrease in volume is enhanced by pressure;
- Principle of microscopic ordering: at constant temperature, an increase in pressure increases the degrees of ordering of molecules of a given substance. therefore pressure and temperature exert antagonistic forces on molecular structure and chemical reactions;
- Isostatic principle: the principle of isostatic processing is presented in the food products are compressed by uniform pressure from every direction and then returned to their original shape when the pressure is released. the products are compressed independently of the product size and geometry because transmission of pressure to the core is not mass/time dependant thus the process is minimized.

If a food product contains sufficient moisture, pressure will not damage the product at the macroscopic levels as long as the pressure is applied uniformly in all directions.

High Pressure Equipment

Even if the principles of HPP and its influence on microbial inactivation have been recognized since late 1800's, the first commercial installation for HPP appeared in Japan in 1990. Although high pressure technology is currently more expensive than traditional processing technologies, the use of high pressure offers new opportunities for food industry to respond to the demand from consumers.

A high-pressure system consists of a high-pressure vessel and its closure(s) pressure-generation system, temperature- control device and material-handling system. the pressure vessel is the most important component of high- hydrostatic-pressure equipment. Several aspects must be taken into account in vessel design. it is necessary to design the high-pressure vessel to be dimensionally stable in a safe- fail way. if it fails it should fail with leak before fracture. Pressure-transmitting fluids are used in the vessel to transmit pressure uniformly and instantaneously to the products sample. Most widely used fluids are water, glycol solutions, silicone oil, sodium benzoate solutions, ethanol solutions, inert gases and castor oil. the food products should be packaged in a flexible packaging. The packages are loaded into the high pressure chamber. the vessel is sealed and the vessel filled with pressure transmitting agent. The high pressure is usually carried out with water as a hydraulic fluid to facilitate the operation and compatibility with food materials. the basis for applying high pressure to foods is to compress the water surrounding the food. At room temperature, the volume of water decreases with an increase in pressure. Because liquid compression results in a small volume change, high-pressure vessels using water do not present the same operating hazards as vessels using compressed gases. ones the desired pressure is reached the pump or piston is stopped, the valves are closed and the pressure can be maintained without further energy input. After holding the product for the desired time at the target pressure, the vessel is decompressed by releasing the pressure-transmitting fluid. For most applications, products are held for 3-5 min at 600 MPa. Approximately 5-6 cycles per hour are possible, allowing time for compression, holding, de-compression, loading and unloading. After pressure treatment, the processed product is removed from the vessel and stored in a conventional way.

High pressures can be generated by direct or indirect compression or by heating the pressure fluid.

Direct Compression

It is generated by pressurizing a fluid by a piston, driven at its larger diameter end by a low pressure pump. This method allows very fast compression, but the limitations of the high-pressure dynamic seal between the piston and the vessel's internal surface restrict the use of this method to small-diameter laboratory or pilot plant systems.

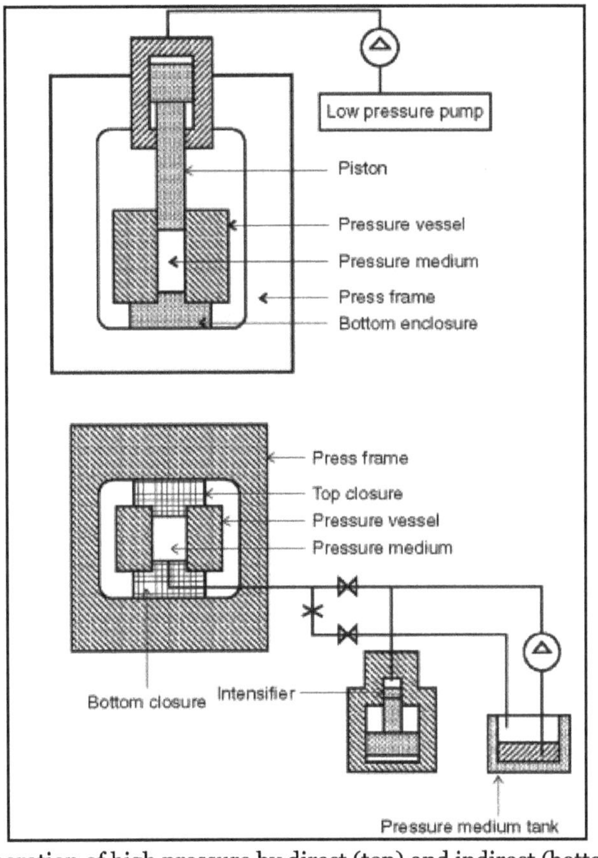

Generation of high pressure by direct (top) and indirect (bottom) compression of the pressure-transmitting medium.

Indirect Compression

This technique uses a high-pressure intensifier to pump a pressure medium from a reservoir into a closed high-pressure vessel until the desired pressure is reached.

Heating of the Pressure Medium

It utilizes expansion of the pressure fluid with increasing temperature to generate high pressure. this method is therefore used when high pressure is applied in combination with high temperature and requires very accurate temperature control within the entire internal volume of the pressure vessel.

Pressure-temperature Effect

to understand and foresee the effect of HPP on foods it is necessary to take in attention the net combined pressure- temperature effect on the treated foods.

During the compression phase (t_1 –t_2) of pressure treatment food products undergo a decrease in volume as a function of the pressure. the product is held under pressure for a certain time (t_2–t_3) before decompression (t_3–t_4). Upon decompression, the product will usually expand back to its initial volume. The compression and decompression can result in a transient temperature change in the product during treatment. The temperature of food (T_1-T_2) increases as a result of physical compression (P_1-P_2). Product temperature (T_2–T_3) at process pressure (P_2-P_3) is independent of compression rate as long as heat exchange between the product and the surroundings is negligible. In a perfectly insulated (adiabatic) system, the product will return to its initial temperature upon decompression (P_3-P_4). In practice, however, the product will return to a temperature (T_4) slightly lower than its initial temperature (T_1) as a result of heat losses during the compression phase. The rapid heating and cooling resulting from HPP treatment offer a unique way to increase the temperature of the product only during the treatment and to cool it rapidly thereafter. The temperature of water increases about 3°C for every 100 MPa of increased pressure at room temperature. On the other hand, fats and oils have a heat of compression value of 8-9°C/100 MPa, and proteins and carbohydrates have intermediate heat of compression values (41, 44).

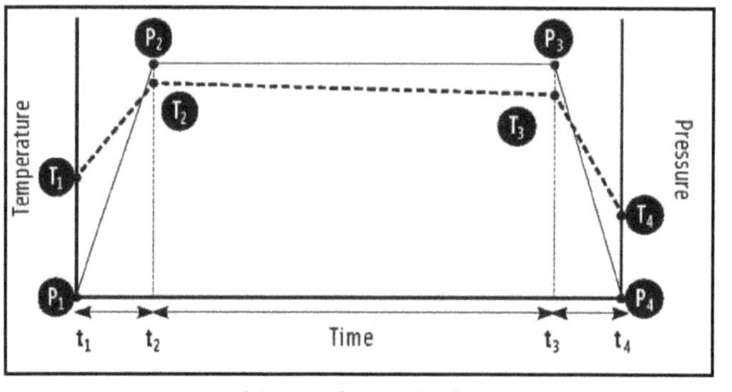

Pressure-temperature history during high pressure processing.

High Pressure Processing and Microbial Inactivation

Microbial inactivation is one of the main goals for the application of high pressure technology.

The extent of microbial inactivation that is achieved by suitable high pressure treatment depends on a number of interacting factors, including type and number of microorganisms, magnitude and duration of high pressure treatments, temperature and composition of the suspension media or food.

The pressure sensitivity of microorganisms may vary between species and probably among the strains of the same species. Microorganisms can be divided into those that are relatively pressure sensitive and those that are pressure resistant. Generally Gram-positive bacteria are more resistant to pressure than Gram-negative bacteria, moulds and yeasts. Most resistant to high hydrostatic pressure are bacterial spores. the sensitivity of microbial cells depends on the stage of the growth cycle at which the organisms are subjected to high hydrostatic pressure treatment. in general, cells in the exponential phase are more sensitive to pressure treatments than cells in the log or stationary phases of growth (16, 42). the grater resistance to pressure when the cell metabolism is slowed down may be due to accumulation of cell components that reduce the effect of high pressure. Rich nutrient media such as meat reinforce the resistance of the microorganisms to HPP. Carbohydrates, proteins and lipids also have a protective effect. A low water activity protects microorganisms against pressure and tends to inhibit pressure inactivation with noticeable retardation as water activity falls.

The extent and duration of high pressure treatment influence the microbial inactivation. An increase in pressure increases microbial inactivation. however, the duration of treatment is increased that does not necessary leads to an increase in the lethal effect. As mentioned above, the microbial response to high-pressure treatments depends on the type of microorganism. For each of them, there is a pressure-level threshold beyond which no effects are detected by increasing the exposure time. There also exists a pressure level at which increasing the treatment time causes significant reductions in the initially inoculated microbial counts.

The temperature during high pressure treatment influences the range of inactivation of microbial cells. Some authors showed that the pressure resistance of microbial organisms is maximal at temperature 15-30°C and decreases significantly at higher or lower temperature.

The treatment of microbial cell and spores with high pressure results in many changes in the morphology, cell membranes, biochemical aspects and genetic mechanisms and all these processes are related to the inactivation of microorganisms. the lethal effect of high pressure on vegetative microorganisms is thought to be the result of a number of possible changes that take place simultaneously in the microbial cell.

The membrane is the most probable site of disruption. high-hydrostatic pressure treatments can alter membrane functionalities such as active transport or passive permeability and therefore perturb the physicochemical balance of the cell. there is a considerably evidence that pressure tends to loosen the contact between attached enzymes and membrane surface as a consequence of the changes in the physical state of lipids that control enzyme activity. the leakage of intracellular constituent through the permeabilized cell membrane is the most direct reason for cell death after high pressure treatment.

Inactivation of key enzymes, including those involved in DNA replication and transcription is also mentioned as a possible inactivating mechanism.

Bacterial spores have demonstrated pressure resistance and the mechanisms through which they are inactivated are different from these for the vegetative cells. it has been suggested that the spore proteins are protected against solvation and ionization. Microbial spores could be inactivated by chosen suitable conditions for high pressure treatment: higher pressure and long exposure time at elevated temperature. it was assumed that pressure caused inactivation of spores by first initiating germination and then inactivating germinated forms. the spore germination could be induced by hydrostatic pressure of 100-300 MPa and resultant vegetative cells are sensitive to environmental conditions.

Usually for pasteurization purpose the considered treatment is generally in the range of 300-600 MPa for a short period of time, from seconds to minutes, inactivating vegetative pathogenic and spoilage microorganisms. For sterilization the range is over 600 MPa and combination with high temperature is needed because some spores are resistant even to pressure over than 1000 MPa when the temperature is not higher than 45-75°C. Most yeast are inactivated by exposure to 300-400 MPa at 25°C within a few minutes, however, yeast ascospores may require treatment at higher pressure. Pressure inactivation of moulds follows a model similar to yeast.

To explain the response of microorganisms to different pressure, high-pressure effects on several biological molecules have been studied. Protein denaturation, lipid phase change and enzyme inactivation can perturb the cell morphology, genetic mechanisms, and biochemical reactions. however, the mechanisms that damage the cells are still not fully understood.

Aspects of Applications of High Pressure Processing of Foods

High-pressure processing provides a unique opportunity for food processors to develop a new generation of value-added food products having superior quality and shelf-life to those produced conventionally.

High pressure processing is a very promising technology for ready-to-eat (Rte) meats because there are few barriers to approval by regulatory authorities, no special labelling requirements because no chemicals are added, and if used appropriately there are no changes to texture or flavour of the product. Researchers found that in Rte meats that are pressure treated at 600 MPa at 20°c for 180 sec, there were no changes in sensory quality, no difference in consumer acceptability, a 4 log reduction in Listeria monocytogenes in inoculated product and the refrigerated shelf-life was extended. There is report that hPP treatment (600 MPa for 10 minutes at 30°c) could extend the shelf-life of food including cooked ham, dry cured ham and marinated beef loins. high pressure application of 500 MPa could extends the shelf life of cooked pork ham and raw smoked pork loin up to 8 weeks, ensuring good microbiological and sensory quality of products.

High pressure processing can effectively inactivate the spoilage microorganisms of several foods, and important food- borne pathogens such as Campylobacter jejuni, Escherichia coli o157:h7, Listeria monocytogenes, and Salmonella spp. Foods can be pasteurized at low or moderate temperatures under pressure. Pressurization at high temperature can sterilize foods. Pressure treatment is of special interest for products or meals containing ingredients that are extensively modified by heat.

HPP has potential as a phyto sanitary treatment to control quarantine insect pests in fresh or minimal processed fruits and vegetables to extend their shelf-life (7, 38). Pressure inactivation of yeast and moulds has been reported in citrus juices. Juices pressurized at 400 MPa for 10 min at 40°c did not spoil during 2-3 months of storage. the high pressure treatment effectively reduced the bacterial flora of fresh goat milk cheese and significantly extended the refrigeration storage life no surviving E. coli was detected in cheese after 60 days of storage (2-4°C) in inoculation studies after treatments at 400-500 MPa for 5-10 min.

High-pressure technology could improve the microbiological safety and quality of foods, including meat, milk, and dairy products.

Biopreservation

Biopreservation can be defined as the extension of shelf life and food safety by the use of natural or controlled micro biota and/or their antimicrobial compounds. One of the most common forms of food bio preservation is fermentation, a process based on the growth of microorganisms in foods, whether natural or added. These organisms mainly comprise lactic acid bacteria, which produce organic acids and other compounds that, in addition to antimicrobial properties, also confer unique flavours and textures to food products. Traditionally, a great number of foods have been protected against spoiling by natural processes of fermentation. Currently, fermented foods are increasing in popularity (60% of the diet in industrialized countries) and, to assure the homogeneity, quality, and safety of products, they are produced by the intentional application in raw foods of different microbial systems (starter/protective cultures). Moreover, because of the improved organoleptic qualities of traditional fermented food, extensive research on its microbial biodiversity has been carried out with the goal of reproducing these qualities, which are attributed to native micro biota, in a controlled environment.

The starter cultures of fermented foods can be defined as preparations of one or several systems of microorganisms that are applied to initiate the process of fermentation during food manufacture fundamentally in the dairy industry and, currently, extended to other fermented foods such as meat, spirits, vegetable products, and juices. The bacteria used are selected depending on food type with the aim of positively affecting the physical, chemical, and biological composition of foods, providing attractive flavour properties for the consumer. To be used as starter cultures, microorganisms must fulfill the standards of GRAS status (Generally Recognized as Safe by people and the scientific community) and present no pathogenic nor toxigenic potential. In addition, use must be standardized and reproducible. The same cultures have been employed for different uses and under different conditions. For the starter cultures, generally LAB, metabolic activity, such as acid production in cheese, is of great technological importance, whereas antimicrobial activity is secondary. However, for the protective culture, generally LAB also, the objectives are the opposite and must always take into account an additional factor for safety as its implantation must reduce the risk of growth and survival of pathogenic microorganism. An ideal strain would fulfil both the metabolic and antimicrobial traits.

Lactic Acid Bacteria

LAB include the genera Lactococcus, Streptococcus, Lactobacillus, Pediococcus, Leuconostoc, Enterococcus, Carnobacterium, Aerococcus, Oenococcus, Tetragenococcus, Vagococcus, and Weisella They form a natural group of Gram-positive, nonmotile, non-sporeforming, rod- and coccus-shaped organisms that can ferment carbohydrates to form chiefly lactic acid; they also have low proportions of G+C in their DNA (< 55%). LAB present attractive physiological properties and technological applications (resistance to bacteriophages, proteolytic activity, lactose and citrate fermentation, production of polysaccharides, high resistance to freezing and lyophilization, capacity for adhesion and colonization of the digestive mucosa, and production of antimicrobial substances).

In general, LAB have GRAS status and play an essential role in food fermentation given that a wide variety of strains are employed as starter cultures (or protective cultures) in the manufacture of dairy, meat, and vegetable products. The most important contribution of these microorganisms is the preservation of the nutritional qualities of the raw material through extended shelf life and the inhibition of spoilage and pathogenic bacteria. This contribution is due to competition for nutrients and the presence of inhibitor agents produced, including organic acids, hydrogen peroxide, and bacteriocins There are many reviews on reported examples of spoilage and pathogenic bacteria inhibition by bacteriocin-producing LAB.

In addition to the food applications of LAB, various strains are considered to be probiotics. Probiotics can be described as a preparation of or a product containing viable, defined microorganisms in sufficient numbers to alter the microbiota (by implantation or colonization) in a compartment of the host and that exert beneficial health effects in this host. In this regard, LAB fit many of requirements for a microorganism to be defined as an effective probiotic. These requirements include the ability to: (a) adhere to cells; (b) exclude or reduce pathogenic adherence; (c) persist and multiply; (d) produce acids, hydrogen peroxide, and bacteriocins antagonistic to pathogen growth; (e) be safe, noninvasive, noncarcinogenic, and nonpathogenic; and (f) coaggregate to form a normal balanced flora. Strains that are used as probiotics for man have been isolated from the human gastrointestinal tract and usually belong to species of the genera Lactobacillus and Bifidobacterium. However, strains belonging to species of other LAB have been used in the past as probiotics as well, such as E. faecium, E. faecalis, S. thermophilus, L. lactis subsp. lactis, Le. mesenteroides, and P. acidilactici.

LAB Bacteriocins

The antimicrobial ribosomally synthesized peptides produced by bacteria, including members of the LAB, are called bacteriocins. Such peptides are produced by many, if not all, bacterial species and kill closely related microorganisms. Due to their nature, they are inactivated by proteases in the gastrointestinal tract. Most of the LAB bacteriocins identified so far are thermostable cationic molecules that have up to 60 amino acid residues and hydrophobic patches. Electrostatic interactions with negatively charged phosphate groups on target cell membranes are thought to contribute to the initial binding, forming pores and killing the cells after causing lethal damage and autolysin activation to digest the cellular wall.

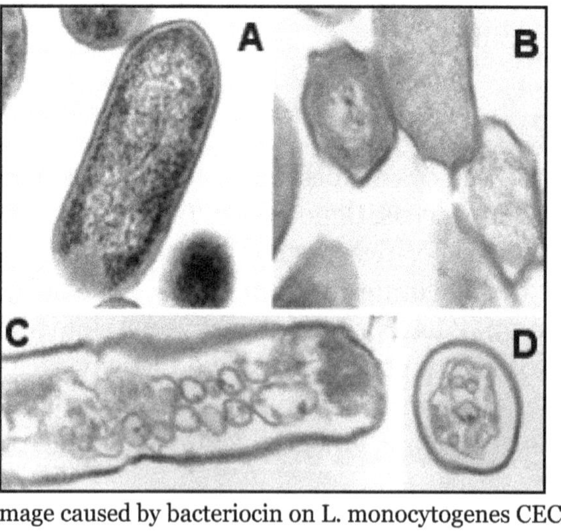

Example of damage caused by bacteriocin on L. monocytogenes CECT 4032 cells.

The LAB bacteriocins have many attractive characteristics that make them suitable candidates for use as food preservatives, such as:

- Protein nature, inactivation by proteolytic enzymes of gastrointestinal tract;
- Non-toxic to laboratory animals tested and generally non-immunogenic;
- Inactive against eukaryotic cells;

- Generally thermoresistant (can maintain antimicrobial activity after pasteurization and sterilization);
- Broad bactericidal activity affecting most of the Gram-positive bacteria and some, damaged, Gram-negative bacteria including various pathogens such as L. monocytogenes, Bacillus cereus, S. aureus, and Salmonella;
- Genetic determinants generally located in plasmid, which facilitates genetic manipulation to increase the variety of natural peptide analogues with desirable characteristics.

For these reasons, the use of bacteriocins has, in recent years, attracted considerable interest for use as biopreservatives in food, which has led to the discovery of an ever-increasing potential of these peptides. Undoubtedly, the most extensively studied bacteriocin is nisin, which has gained widespread applications in the food industry. This FDA-approved bacteriocin is produced by the GRAS microorganism Lactococcus lactis and is used as a food additive in at least 48 countries, particularly in processed cheese, dairy products and canned foods. Nisin is effective against food-borne pathogens such as L. monocytogenes and many other Gram-positive spoilage microorganisms. Nisin is listed in Spain as E-234, and may also be cited as nisin preservative or natural preservative. In addition to the work on nisin, several authors have outlined issues involved in the approval of new bacteriocins for food use.

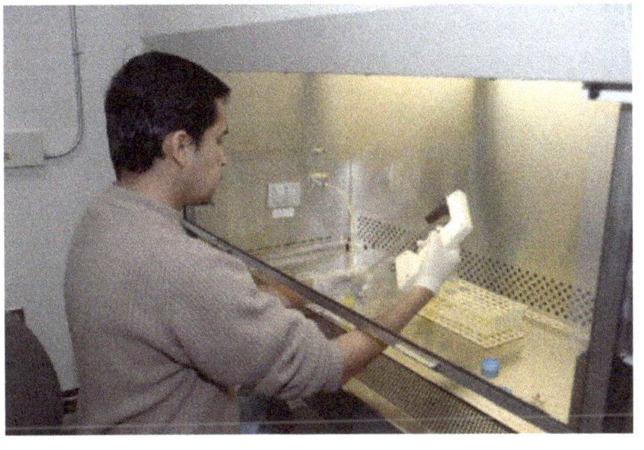

Hurdle Technology

"Hurdle Technology" is a technology that uses a combination of two or more preservation parameters at an optimum level in order to get a maximum lethality against micro-organisms without compromising with the nutritional and sensory qualities of food products. A wide range of preservation techniques is available e.g. freezing, blanching, pasteurizing and canning but the spoilage and poisoning of foods by microorganisms is a problem that is not yet under adequate control. Moreover, today's consumer demands for more natural and fresh-like foods, that requires the use of only mild preservation techniques. Hence, for the benefit of food manufacturers and fulfill the requirement of the consumer, there is a strong need for new or improved mild preservation methods that allow for the production of fresh-like, but stable and safe food. This topic would be primarily focussing on hurdle technology in food preservation.

Hurdle Technology in Food Preservation.

Types of Hurdles

There are more than 60 potential hurdles those can be used for food preservation but the most important hurdles used are:

1. Temperature(high or low),
2. Water Activity (aw),
3. Acidity(ph),
4. Redox Potential (Eh),
5. Chemical Preservatives (Ex-nitrite, sorbate, sulfide),
6. Competitive Microorganisms (Ex-lactic acid bacteria).

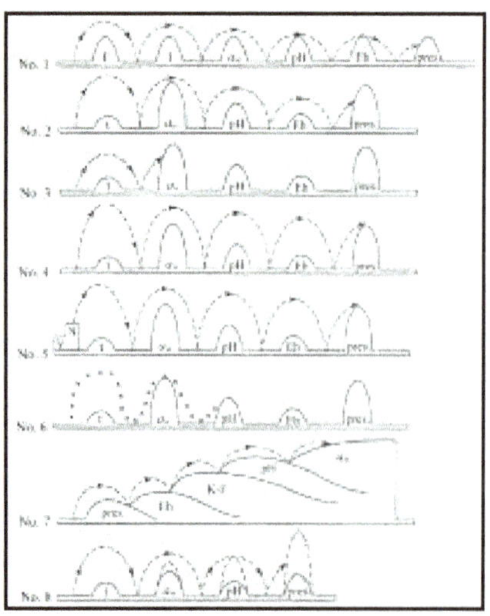

The preservation of almost all the foods is a result of combined application of several preservative methods (e.g., heating, chilling, drying, curing, conserving, acidification, oxygen-removal, fermenting, adding preservatives, etc.). These methods and their underlying principles (i.e., F, t, aw, pH, Eh, competitive flora, preservatives, etc.) have been applied since long back in our traditional foods empirically, but as the knowledge about these factors has increased, they are being applied intelligently using the concept of hurdle technology. Recently several novel preservative techniques (e.g. gas packaging, biopreservation, bacteriocins, ultra high-pressure treatment, edible coatings, etc.) have gained popularity to be used in combination with other traditional preservative factors(Hurdles).

Every hurdle could have both, positive or a negative effect on foods, depending on its intensity. For example, use of low temperature (chilling) below the critical limit of any food can lead to "chilling injury" whereas moderate chilling will be beneficial for their shelf life extension as it retards microbial growth. Similarly lowering the pH in fermented sausage inhibits the growth of pathogenic bacteria but lowering beyond the required limit can also impair the taste. Therefore a balanced intensity of any hurdle should be used for food preservation. The Too small intensity of hurdle should be strengthened and if it is detrimental to food quality, it should be lowered. This adjustment will provide a food that would be safe as well as will have a good quality by the use of hurdles in an optimal range.

Mechanism of Food Preservation by Hurdle Technology

The whole mechanism of preservation of food by using the concept of hurdle technology is comprised of various responses those are given by any microorganism. The whole phenomenon can be understood by following:

- Homeostasis: Tendency of any organism to maintain its internal status is known as Homeostasis. The homeostasis of microorganisms plays a key role in food preservation. If any of the hurdles used in food disturbs the homeostasis of these microorganisms, they will not be able to multiply and will remain constant in number or will die before the re-establishment of homeostasis. Therefore, food preservation can be achieved by disturbing the homeostasis of microorganism, temporarily or permanently;
- Metabolic Exhaustion: Another important phenomenon for food preservation is Metabolic Exhaustion of microorganisms. As a response to the hurdles applied to foods, microorganisms try to repair their homeostasis, use up all their energy for this and become metabolically exhausted. This leads to an auto-sterilization of such foods. The foods which are preserved with the concept of hurdle technology and are microbiologically stable, become safer during storage at ambient temperature. The microbes can respond better to the hurdles at ambient temperature than at refrigeration and become metabolically exhausted;
- Stress Reactions: As a response of various hurdles e.g. heat, pH, aw , ethanol, oxidative compounds, etc. as well starvation, a stress shock protein is generated by some bacteria. These stress proteins may hamper food preservation and could turn out to be problematic for the application of hurdle technology if only one hurdle has been applied. If different stresses are received by the microorganism at the same time, the activation of genes for the synthesis of stress shock proteins, which help organisms to cope with stress situation, would be difficult. Synthesis of many stress shock proteins due to simultaneous exposure to different stresses will be very energy-consuming and would lead to metabolic exhaustion of the microorganism;
- Multi-Target Preservation: A combined effect could be achieved by hitting various targets within the microbial cell (e.g., cell membrane, DNA, enzyme systems, pH, aw, Eh) by using

different hurdles simultaneously. This disturbs the homeostasis of the microorganisms present in several respects. In this case, the replenishment of homeostasis and activation of stress shock proteins becomes more difficult. Therefore, simultaneous application of different hurdles in a particular food would lead to optimal microbial stability.

References

- "Food Irradiation in Asia, the European Union, and the United States" (PDF). Japan Radioisotope Association. May 2013. Archived from the original (PDF) on February 9, 2015. Retrieved 26, April 2020
- Food-preservation-methods-canning-freezing-and-drying, canning, food-drink: dummies.com, Retrieved 30 July 2020
- Fan Xuetong (2011). "Changes in Quality, Liking, and Purchase Intent of Irradiated Fresh-Cut Spinach during Storage". Journal of Food Science. 76: S363–S368. doi:10.1111/j.1750-3841.2011.02207.x
- Rennie, Richard (2016). boiling-point elevation. A Dictionary of Chemistry. Oxford: Oxford University Press. ISBN 9780198722823
- Salting-pickling-processes-food-preparation-preservation: foodsafetyhelpline.com, Retrieved 18 March 2020
- Rickus, Alexis; Saunder, Bev; Mackey, Yvonne (2016-08-22). AQA GCSE Food Preparation and Nutrition. Hodder Education. ISBN 9781471863653
- Publishing, D. K. (2005-08-29). The Cook's Book: Techniques and tips from the world's master chefs. Penguin. ISBN 9780756665609
- Bhat R.; Sridhar K. R.; Tomita Y.; Tomita-Yokotanib K. (2007). "Effect of ionizing radiation on antinutritional features of velvet bean seeds (Mucuna pruriens)". Food Chemistry. 103: 860–866. doi:10.1016/j.foodchem.2006.09.037

Chapter 4

Chemical Preservation in Food Processing

The process of adding chemicals to food products to prevent the deterioration of food caused by the growth of micro-organisms is known as chemical preservation. Some of the chemical preservatives are sodium benzoate, nitrites and nitrates, sulfites, potassium sorbate and spices. This chapter has been carefully written to provide an easy understanding of the varied facets of chemical preservation in food processing.

Preservative for food may be defined as any chemical compound and/or process, when applied to food, retard alterations caused by the growth of microorganisms or enable the physical properties, chemical composition and nutritive value to remain unaffected by microbial growth. Some chemicals have been used traditionally since several decades as direct or indirect inhibitors of microbial growth and are still widely used despite their limitations.

The majority of food preservation operations used today also employ some kind of chemical additive to reduce spoilage. Of the many dozens of chemical additives available, all are designed either to kill or retard the growth of pathogens or to prevent or retard chemical reactions that result in the oxidation of foods.

Some familiar examples of the former class of food additives are sodium benzoate and benzoic acid; calcium, sodium propionate, and propionic acid; calcium, potassium, sodium sorbate, and sorbic acid; and sodium and potassium sulfite. Examples of the latter class of additives include calcium, sodium ascorbate, and ascorbic acid (vitamin C); butylated hydroxyanisole (BHA) and butylated hydroxytoluene (BHT); lecithin; and sodium and potassium sulfite and sulphur dioxide.

There are several traditional methods of food preservation used at the household level that can be classed as chemical methods. Substances such a sugar, salt, vinegar, spices and wood-smoke are generally regarded as safe and natural preservatives. Salting, sugaring and smoking are all methods of curing foods. Curing is a general term that covers all these types of food preservation.

Salting is the addition of salt (sodium chloride or NaCl) to food for the purpose of preservation. The growth of microorganisms is inhibited by the salt, which has the effect of drawing water out of the bacterial cells so they become dehydrated and die. In this manner, salt, in combination with other measures, acts as a preservative in many foods such as butter, cabbage, cheese, cucumber, meat and fish. It also gives a desired flavour to the food. Salting can be done by rubbing adequate quantities of dry salt into foods, or by immersion, where the food item is soaked in a concentrated salt solution (i.e. brine). For effective preservation, the concentration of the brine solution has to be maintained above 18%. This is approximately one cupful of salt to five cups of water.

Sugaring refers to the action of sugar in food preservation. It is similar to the action of salt in that it depends on the removal of water. In concentrations of at least 65%, sugar solution is widely used as a sweetening and preserving agent. However, care is needed because at low concentrations, sugar solution can support the growth of microorganisms. It has been found that microorganisms rarely survive in solutions above 20–25% sugar concentration.

Smoking is one of the oldest methods used to improve the quality of food and is commonly used to preserve meat and fish. The smoking process involves exposing food to smoke from burning or smouldering wood or other plant material. It partially preserves the food by surface drying, i.e. removing moisture from the surface of the food, but it is not a reliable method of preservation unless combined with some other method such as salting or drying.

Spices also have some uses in food preservation because they tend to inhibit the growth of staphylococci and other bacteria. However, they have a very limited application because they often get contaminated themselves by a number of bacteria.

Sodium Benzoate

Sodium benzoate is a common type of food preservative and is the sodium salt of benzoic acid. Food manufacturers make sodium benzoate by synthesizing the compounds, sodium hydroxide and benzoic acid, together. In addition to its use as a food preservative, sodium benzoate has other roles in food production as well. There are some side effects associated with excess sodium benzoate consumption in foods, so talk with your doctor about it to make sure you are not ingesting harmful doses of preservatives.

Sodium benzoate preserves food by having anti-fungal properties, protecting foods from invasion by fungi that cause food to spoil and potentially make you sick. Sodium benzoate works by entering the individual cells in the food and balancing its pH level, increasing the overall acidity of the food. By lowering the intracellular pH of certain foods, sodium benzoate creates an environment in which fungi cannot grow and spread.

Soft Drinks and Acidity

Sodium benzoate is heavily used by the soft drink industry due to the demand of high-fructose corn syrup in carbonated drinks. Sodium benzoate increases the acidity of soft drinks, which also increases the intensity of flavour you get from the high-fructose corn syrup. On the back of a soda can, you can find sodium benzoate in the ingredients list as E211, which is the number assigned to it as a food additive.

Sodium benzoate is also used in medicine preparations, dye manufacturing, tobacco products and as a rust and mildew inhibitor. Sodium benzoate in food can be toxic when combined with other foods containing vitamin C.

Acidic Food Flavour Enhancement

Sodium benzoate is primarily added to acidic foods to enhance their flavor. It can be found in foods such as pickles, sauces, jams and fruit juices. Foods that contain vinegar, such as salad dressings, typically contain very high levels of sodium benzoate. Benzene, a precursor to sodium benzoate, can be found in very small amounts naturally in some fruits, vegetables, meats, dairy products and even drinking water.

Harmful Side Effects

Sodium benzoate, when combined with vitamin C, forms benzene. This substance is a carcinogen and is known to contribute to the formation of many different types of cancer. Sodium benzoate and vitamin C can occur together in beverages such as fruit juice or soda, where these substances have been added to prevent spoilage and inhibit growth of bacteria, molds and yeast. Benzene can form when beverages containing both substances are exposed to heat or light.

However, the Food and Drug Administration states that food products that contain both vitamin C and sodium benzoate express benzene levels that are below the dangerous limit. Talk to your doctor about sodium benzoate, especially if you consume a lot of foods and drinks that contain high levels of this food additive.

Nitrites and Nitrates

The salts of nitrite and nitrate are commonly used for curing meat and other perishable produce. They are added to food to preserve it and also help hinder the growth of harmful microorganisms, in particular Clostridium botulinum, the bacterium responsible for life-threatening botulism. Nitrites, together with nitrates, are also added to meat to keep it red and give flavour, while nitrates are used to prevent certain cheeses from bloating during fermentation. Nitrate is found naturally in vegetables, with the highest concentrations occurring in leafy vegetables like spinach and lettuce. It can also enter the food chain as an environmental contaminant in water, due to its use in intensive farming methods, livestock production and sewage discharge.

What happens to Nitrites/Nitrates in the Body?

In humans, nitrite and nitrate from food are rapidly absorbed by the body and, for the most part, excreted as nitrate. Some of the nitrate absorbed by the body is recirculated through salivary glands and part of it is converted by mouth bacteria into nitrite. Absorbed nitrite can oxidise haemoglobin to methaemoglobin, an excess of which reduces the ability of red blood cells to bind and transport oxygen through the body. Nitrite in food (and nitrate converted to nitrite in the body) may also contribute to the formation of a group of compounds known as nitrosamines, some of which are carcinogenic.

Food Sources of Nitrates and Nitrites

The total nitrates intake in humans mainly derives from food (80%) and water (14%). Although the different lifestyles and dietary habits in various countries can lead to wide variations in the amount of consumed nitrates, the World Health Organization estimates that the mean daily dietary intake of nitrates is usually ranged from 43 to 141 mg.

Several works indicate that the majority of the nitrates intake is attributed to plant foods. Approximately 98% of the nitrates intake from food is attributed to fruits and vegetables, while only the remaining 2% to cured meat products. 87% of the dietary nitrates intake is associated with the consumption of vegetables. Several factors (e.g. nitrates uptake, nitrates reductase activity, growth rate and growth conditions, processing methods such as heat treatments and storage conditions) affect the ultimate nitrates content of vegetables. A factor of great importance is the application of fertilizers generally resulting in a greater uptake of nitrogen in vegetables and eventually in higher nitrates content. Leafy vegetables such as lettuce and spinach tend to have higher levels of nitrates than seeds or tubers.

In contrast to nitrates, nitrites are mainly formed from the endogenous nitrates conversion. Dietary nitrites represent less than 20 % of the total daily nitrites exposure, the remaining 80 % resulting from the endogenous bioconversion of dietary nitrates to nitrites in saliva. Humans generally consume 1.2– 3.0 mg of nitrites per day. Among consumed food, the total daily nitrites intake may be as high as 4% and 19% for fresh meat and cured meat, respectively. Vegetables represent only 19% of the total daily nitrites intake. Cured meat products comprise 4.8% of daily nitrites intake and vegetables account for just 2.2%.

The Nitrates and Nitrites Changes in Cured Meat Products

The nitrates are inert compounds and should be reduced to nitrites, which are the highly active curing compounds. The reduction of nitrates to nitrites is mainly accomplished by either naturally present bacteria in the meat or by the addition of bacteria possessing a nitrates reductase activity. The most efficient nitrates reducing organisms are staphylococci and micrococci, which also affect the fermentation process in meat products. Lactic acid bacteria can also reduce nitrates to nitrites but their potential has not been used in commercial starter cultures. Although it is generally accepted that the nitrates reduction is exclusively performed by microorganisms, nitrates reductase activity is normally present in animal physiology. Furthermore, Nitrates reduction has been also observed in heated cured meat products, which may be due to chemical reactions.

The rate of nitrates reduction varies in different meat products. In certain traditional products, nitrates act as a reservoir of nitrites, the rate of reduction being dependent on the nature of the natural microbial flora. Occasionally nitrates can be also detected in meat products processed with nitrites additives, as the result of microbial activity.

In meat, nitrites are usually decreased rapidly and converted to a variety of compounds such as nitrous acid, nitric oxide (NO) and nitrates. At a pH of the meat, generally 5.6 to 5.8, nitrites are mostly present in the dissociated form, and are converted to several intermediate compounds, which are unstable and therefore difficult to be determined. The compounds derived from nitrites can act as oxidizing, reducing or nitrosylating agents.

The decrease of nitrites in meat products may be due to the action of meat endogenous reducing substances, e.g. sulphur containing amino acids (cysteine), or to additives such as ascorbate. In the last decades, ascorbic acid or ascorbate, isoascorbate (erythorbate) have been used as additives in cured meat batters. The ascorbate reacts with oxygen, by forming dehydroascorbate, and reduces the amount of nitrites which could be oxidized to nitrates. Ascorbate seems also to react with resulting NO; the so formed new NO compounds can further react with other meat ingredients.

It is generally accepted that the formation of NO from nitrites is a prerequisite step for most meat curing reactions. NO reacts in various ways, concurrent with one another. The decrease in the NO amount is due to NO reactions with myoglobin and other substrates present in meat, e.g. amino acids such as cysteine or glutathione.

Consequently many of the nitrites added in the meat are decreased through a series of NO reactions during manufacture and storage of cured meat products. The rate of nitrites loss depends on many factors such as the temperature of the heat process, the pH of the product, the storage temperature and the addition of ascorbic acid or other reducing agents. An amount of 10% to 20% of the originally added nitrites is typically present after the manufacturing process. This amount of nitrites, referred to as residual nitrites, slowly decline over the storage life of the cured meat products, and sometimes reach non-detectable levels. In most meat products, there is no simple and direct relationship between the initial amount and the residual nitrites amount.

Beneficial Effects of Nitrates or Nitrites Addition in Meat Products

The use of nitrates and nitrites as curing agents has beneficial effects on the quality factors as well as on the microbiological.

Quality Effect

The addition of nitrates and nitrites in cured meat shows a positive effect on quality meat characteristics such as color, flavor or stability against oxidation. The color of fresh meat is red to shaded pink, which is an extremely important attribute for consumer acceptance. The basic color component of meat is myoglobin, which is a globular protein consisting of an heme group with a centrally located atom of iron (Fe) surrounded by a globin moiety. According to the iron oxidation state (Fe^{2+} or Fe^{3+}) the myoglobin may be found in muscles in three states (myoglobin, oxymyglobin and metmyoglobin). Originally, myoglobin (Fe^{2+}) does not bind any ligand. After slaughtering of the animal, the oxygen in direct contact with the muscles can bind Fe^{2+} of myoglobin and form oxymyglobin, which is red in color but not very stable. However, Fe^{2+} may further oxidized to Fe^{3+} by oxidizing compounds such as oxygen or nitrites. Thus, this oxidation in the presence of oxygen leads to formation of metmyoglobin (Fe^{3+}), which is brown in color, and it is also a less desirable attribute to consumers.

In cured meat, the red color is developing in a number of complicated reaction steps until the characteristic curing red pigment, is formed. NO deriving from nitrites, is bound to the iron ion (Fe^{+2}) located in the centre of the porphyrin ring system of myoglobin and forms NO-myoglobin (nitrosomyoglobin), which is colored but not stable. Nitrosomyoglobin can be transformed to the characteristic stable red pigment of cured meat products (nitroso-myochromogen), either by the application of heat in the heat processed meat products or in the low acid environment of fermented meat products. In heat processed meat products, the protein moiety of the nitrosomyoglobin is denatured, but the red NO-porphyrin ring system (nitrosomyochromogen) which is stable, still exists and is found in meat products heated up to 120°C. In low acid fermented meat products, the protein moiety of nitrosomyoglobin is also denatured but the nitroso-myochromogen is not affected and improves the color stability. It is generally accepted that a small amount (10-15 ppm) of residual nitrites is required for the color stability in cured meat products. The role of nitrites in the cured meat flavor is a quite complex procedure and not fully understood. Meat curing practices not only involve sodium nitrite and salt, but also sugar, reducing agents, phosphates and spices, which also contribute in the flavor. However, they are nitrites that are mainly responsible for the development of the distinct flavour of cured meat products. Since nitrites are bound with proteins and lipids, several volatile and not volatile compounds are formed and contribute to the flavor of the cured meat. Binding of nitrites to sulfur containing amino acids of meat proteins results in the production of SH-residues with specific aroma and flavor. Many studies indicated the difficulties of isolating the flavor compounds of cured meat products. Several flavor compounds have been isolated in cured meat products, such as hydrocarbons, alcohols, ketones, furans, pyrazines, and sulphur- and nitrogen- containing heterocyclic compounds.

Nitrites exhibit antioxidant activity, and slow the breakdown of unsaturated fatty acids, as well as the development of secondary oxidation flavor compounds. The antioxidant effect of nitrites is primarily due to oxygen deletion. The NO molecule itself can easily be oxidized to NO_2 in the presence of oxygen resulting to an oxygen sequestering. Due to lack of oxygen, the development of rancidity or a warmed off flavor are retarded in cured meat products.

Moreover, a stable complex between heme-bound iron and nitrites is formed, which inhibits a release of iron ions. Consequently, free iron ions (Fe^{2+}) are not available for the initiation of lipid peroxidation. Furthermore, nitrites are also able to react with free iron ions and form a complex. It is also assumed that nitrites are able to stabilize polyunsaturated fatty acids forming nitro–nitroso derivatives. As a consequence, cell membranes are protected against lipid peroxidation and the stability of cured meat product during storage is enhanced. The addition of nitrites to meat products also leads to lower contents of harmful cholesterol oxides. In addition, nitrites can form various nitroso- and nitrosyl compounds, like S-nitrosocysteine, which exhibit antioxidant properties.

Nitrites at relatively low levels (40 ppm) proved to inhibit warmed off flavor development. Because nitrites are highly effective as antioxidants, synthetic antioxidants such as butylated hydroxyanisole (BHA) and butylated hydroxytoluene (BHT) are not usually used in the majority of the cured meat products.

Microbiological Safety

Nitrites exhibit important bacteriostatic and bacteriocidal activity against several spoilage bacteria as well as foodborne pathogens found in meat products. However, the majority of studies concerning the antimicrobial properties of nitrites have been focused on C. botulinum in several meat products produced in various countries.

Nitrites show antibotulinal activity against both vegetative cells as well as spores of the pathogen in thermally processed meat products. Despite the vast amount of research, the precise nitrites action against C. botulinum or other bacteria has not absolutely clarified in meat products.

The antimicrobial action of nitrites is likely attributed to reactions associated with the generation of NO or nitrous acid. Researcher suggested that the inactivation of iron-sulphur proteins of C. botulinum vegative cells by NO, is the probable mechanism of inhibition of the pathogen in nitrites-cured meat products. The antimicrobial activity of nitrites in cured meat products is also enhanced from other antimicrobial hurdles such as heat treatment, pH, salt, water activity (a_w), redox potential and other curing ingredients. Several studies indicated that nitrites inhibit bacteria more effectively at low pH. Other components, such as ascorbate or isoascorbate / erythorbate and also cysteine, may enhance the antibotulinum efficacy of nitrites in cured meat by sequestering metal ions. Researcher has shown that ascorbate in meat products reduces the toxin production by proteolytic C. botulinum types A and B together with nitrites and NaCl.

Although in most cured meat products, the nitrites are added to prevent the growth and toxin production by C. botulinum, the required amount for this antimicrobial efficacy varies from product to product. According to the opinion of the Scientific Committee on Food of the European Food Safety Authority (EFSA), 50- 100 mg nitrites per kg meat may be sufficient for the majority of cured meat products; however, for other meat products, especially those with a low NaCl content and a prolonged shelf life, the presence of a higher content, as high as 150 mg kg^{-1} of nitrites, is required for efficacy against C. botulinum. As nitrites levels increase, inhibition of C. botulinum growth and toxin production also increases. The nitrites level required for color development (roughly 25 ppm or less) is quite small as compared to this required for C. botulinum control in cured meat.

Antibacterial compounds are also formed as a result of nitrites-related reactions during storage. The contribution of the residual amount of nitrites in the microbiological safety of meat products has been also disputed. For example, in meat products containing ascorbate (or isoascorbate / erythorbate) resulting in more rapid loss of nitrites, the residual nitrites content is very low, and sometimes below the level of detection, although the growth of C. botulinum is still prevented.

Nitrates have no direct activity against C. botulinum. The addition of nitrates is usually required in particular traditional dry meat products, typical of the Mediterranean countries, like long-ripened dry-fermented sausages of high pH and dry-cured ham. In these products nitrates act as reservoirs of nitrites, progressively reduced to nitrites, which contribute not only to sensory properties but also to safety due to their antimicrobial action.

Control measures such as good manufacturing practice (GMP) and hazard analysis critical control point (HACCP) during production of certain meat products could result in reduction of the required levels of nitrates-nitrites, although these products may be not strictly classified as "cured meat products".

The antimicrobial activity of nitrites against food-borne pathogens in cured meat products has been also studied in recent years. Nitrites have been proved to have a stronger antimicrobial activity against Gram positive than Gram negative bacteria. However, the antimicrobial efficacy of nitrites against Gram negative bacteria is also significant.

In an early study, Researchers reported no nitrites efficacy against Staphylococcus in frankfurters. Researchers found a bacteriostatic action on Staphylococcus aureus and they suggested that their results provided evidence that the pathogen may be effectively controlled with 200 ppm of nitrites. The antimicrobial action of nitrites against S. aureus was also reported by researcher.

Researcher found that nitrites were effective for controlling Gram-negative enteric pathogens such as Salmonella and Escherichia coli. Researcher showed the inhibitory activity of nitrites against enterobacteria and Shigatoxin producing Escherichia coli (STEC) in fermented sausages.

Nitrites with sodium lactate and sodium diacetate have been found to act synergistically against Listeria monocytogenes in cured meat products. Researcher reported inhibitory activity of nitrites against L. monocytogenes, E. coli O157:H7, S. aureus, and Bacillus cereus in cured meat products.

Metabolism of Nitrates and Nitrites

The metabolism of nitrates in the human body has been extensively studied. Nitrates undergo a number of metabolic conversions, and are recycled between the saliva and the gut and the bile and the gut. Briefly, once nitrates are ingested, they are quickly absorbed in the upper part of the small intestine in humans. Absorbed nitrates are rapidly transported by the blood and selectively secreted by the salivary glands, and probably other exocrine glands. Researcher reported that approximately 25% of ingested nitrates are secreted in saliva. In the oral cavity, the bacteria reduce approximately 20% of the secreted nitrates into nitrites, and formed nitrites are then swallowed along with the unconverted nitrates.

Healthy adults have a salivary conversion of nitrates to nitrites of normally 5-7% of the total nitrates intake. In the stomach, nitrites are rapidly transformed to nitrous acid, which in turn is decomposed to nitrogen oxides including NO. Most of the absorbed nitrates in the digestive system are ultimately excreted in the urine, but considerable salvage takes place in advance through selective reabsorption from the kidney and the salivary recirculation. In addition to the dietary intake, nitrates are formed endogenously, with NO as the precursor. The main source of endogenous nitrates in mammals is the L-arginine-NO synthase pathway, which is constitutively active in numerous cell types throughout the body. NO is produced from the amino acid L-arginine and molecular oxygen by nitric oxide synthetase (NOS).

Nitrites can be reduced and form NO in human tissues via several pathways by involving haemoglobin, myoglobin, or xanthine oxidoreductase. The generation of NO by these pathways is mainly enhanced during hypoxia and acidosis. The proteins modified by NO contribute to physiological hypoxic signalling, vasodilation, cellular respiration and cellular response to ischaemic stress.

Concerns Associated with Nitrates and Nitrites in Meat Products

The first international evaluation of the risks associated with the ingestion of nitrates and nitrites was conducted by the Joint FAO/WHO Expert Committee on Food Additives (JECFA) in 1961. Since then, research works have clearly shown that nitrates and nitrites have both acute and long-term effects on human health.

Nitrates per se are relatively non-toxic, but nitrates metabolic products such as nitrites, NO and N-nitroso compounds (NOCs), have raised concern over potential adverse health effects. The acute toxicity of nitrites is approximately 10-fold higher than that of nitrates. The lethal oral doses for human beings are established in 80–800 mg nitrates kg^{-1} bw and 33–250 mg nitrites kg^{-1} bw. The high amount of nitrites added to meat products in the past, had caused deaths due to intoxication, in early decades of 20th century in Germany.

The main adverse effect related to acute toxicity of nitrites is methaemoglobinaemia, manifested as cyanosis, usually as a consequence of excessive nitrites exposure. Infants under 3 months of age are more susceptible than older children and adults due to nitrites inhibiting the transport of oxygen to hemoglobin. A case of acute human intoxication to nitrites in a 40 year-old mother and her 9 years old child was recently described in Italy, after the consumption of turkey contaminated with high nitrites levels ranging from 6,000 to 10,000 mg kg^{-1}.

The long-term toxicity of nitrates and nitrites is associated with their potential to form carcinogenic NOCs in the food matrix as well as in the human body. NOCs are formed when various nitrosating agents originating from nitrites interact with the amide compounds either in food or in the stomach. The NOCs can be divided into two classes: the first class is nitrosoamines (NA) and the second class is nitrosoamide-type compounds such as N-nitrosoureas, N-nitrosocarbamates and N-nitrosoguanidines. NOCs of both classes differ considerably in their chemical formation and biological effectiveness.

The NA commonly found in meat products are the volatile substances N-nitrosodimethylamine (NDMA), -piperidine (NPIP), -pyrrolidine (NPYR), -diethylamine (NDEA). NDMA and NDEA are usually found in some thermally cured products.

NA are formed by a nitrosylation of the secondary amines. Also primary amines can be nitrosated, but these products are not stable. Nitrosation of tertiary amines is not possible. The process of nitrosation is very complicated and depends on many factors such as pH, the type of the secondary amine and temperature.

NA formation is favored under the following conditions:

- Secondary amines should be present. In fresh meat, the secondary amines are either not present or are found at very low amounts. Most amines in meat are primary amines derived from α-amino acids. Potential precursors of secondary amines like creatine and creatinine, the free amino acids proline and hydroxyproline, as well as some decarboxylation products can lead to formation of secondary amines during fermentation and storage of meat products.

- The pH must be low (less than 6.55) to form nitrosating agents. The low pH is only important for fermented sausages.

- Thermal processing at high temperatures, usually above 130 °C (formation of NPYR), or long storage at room temperature (formation of NDMA, NPYR). Grilling, roasting or frying of meat products could form NA.

Several epidemiological studies have suggested NA as the etiological agents of various types of cancer in humans. Several NOCs tested in different animals proved to cause cancer.

Sulfites

Sulphites or sulphating agents are a group of sulphur-based compounds that are capable to release sulphur dioxide SO_2, which is the active component that helps to preserve food. Sulphites have been added to foods as preservative agents and other purposes for centuries. Sulphites were used by many ancient cultures such as Romans and Greeks to reduce or prevent spoilage and discoloration of foods, to sanitize wine vessels or as a purifier and disinfectant. Furthermore, naturally occurring sulphites exist especially in fermented products, such as beer and wine. They are produced by yeasts during the fermentation process.

Sulphites have many functional uses in food. They are effective bleaching agents, antimicrobials, oxygen scavengers, reducing agents, and enzyme inhibitors. Their primary function is to act as a preservative or antioxidant to prevent or reduce/retard spoilage and browning during the preparation, storage, and distribution of many foods. Additionally, sulphites are utilized to stabilize product colour and inhibit discoloration, as well as to improve flavour of many foods. Sulphites are also used in the pharmaceutical and cosmetics industries as preservatives of the active ingredient. Thus, in addition to being cheap and convenient, sulphites are extremely versatile, and in many products, serve more than one purpose.

Despite all uses of sulphites in the food industry, sulphites have been related to asthmatic reactions in a small subset of sensitive individuals of the asthmatic population. They have also been related with some anti-nutritional consequences such as degradation of thiamine (vitamin B_1).

Chemistry of Sulphites

Sulphites are a group of compounds characterized by the presence of the sulphite ion SO_3^{2-}. The term sulphites is commonly used to describe the oxo species of sulphur oxidation state that include the ionic and non-ionic forms: SO_2, HSO_3^-, SO_3^{2-}, and $S_2O_5^{2-}$. The concentration of free sulphites in foods correlates with the preservative activity, with sulphite levels in foods and drinks usually expressed as theoretical yields or 'equivalents' of SO_2. The term S(IV) is used to refer to any of the four oxo species of sulphur oxidation. The most common sulphites used by food industry are described in Table.

Table: Most common sulphites used by food industry.

Name	E-number	Chemical formula
Sulphur dioxide	E-220	SO_2
Sodium sulphite	E-221	Na_2SO_3
Sodium hydrogen sulphite (Sodium bisulphite)	E-222	$NaHSO_3$
Sodium metabisulphite	E-223	$Na_2O_5S_2$
Sodium metabisulphite	E-224	$K_2O_5S_2$
Calcium sulphite	E-225	$CaSO_3$
Calcium hydrogen sulphite	E-226	$Ca(HSO_3)_2$
Potassium hydrogen sulphite	E-227	$KHSO_3$

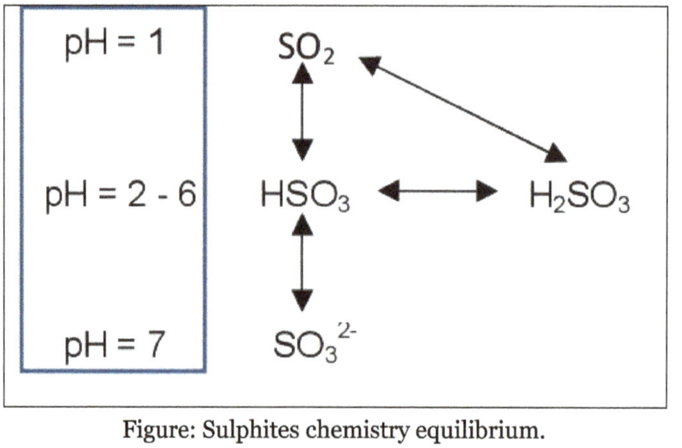

Figure: Sulphites chemistry equilibrium.

A distinction is often made between free and bound S(IV) which forms reversible interaction products. The former refers to all the species of the additive which may rapidly and quantitatively be converted to SO_2, thereby acidifying a treated food. The latter represents hydroxysulphonate adducts formed by reaction of carbonyl groups with HSO_3^-. In fact, according to the nature of food, the type and extent of the technological processes, and the conditions and time of storage, sulphites react with reducing sugars, aldehydes, ketones, proteins to form combined compounds.

In the digestive tract, the free and reversibly bound S(IV) species can be liberated. The useful properties of sulphiting agents are generally due to the nucleophilicity of the sulphite ion that may react by addition to carbonyl groups, carbon-carbon double bonds, quinones, heterocyclic nitrogen compounds. The most common reactions of S(IV) are the result of oxidation by molecular oxygen catalysed by transition metal ions.

Sulphites as Food Additive

Uses and Reactions in Food

Sulphites are used in a wide variety of food products such as beverages, dried fruits, jams, seafood, meat, sugar, and others. In wine, the aim of adding sulphite is to prevent microbial growth and oxidation of volatile odorous compounds which may develop during fermentation.

Table: Major type of foods that may contain sulphiting agents.

Food category	Examples
Drinks	Bottled soft drinks and fruit juice, cordials, cider, beer, wine (including sparkling wine)
Processed fruits	Dried apricots, fruit bars, maraschino cherries, canned, bottled, or frozen fruit juices, commercial preparations of citrus fruit beverage mixes.
Processed vegetables	Canned vegetables (including potatoes), pickled vegetables (including sauerkraut), dried vegetables, instant mashed potatoes, frozen potatoes, potato salad
Fish and crustaceans	Canned clams; fresh, frozen, canned, or dried shrimp; frozen lobster; scallops; dried cod
Other commercial foods	Gravies, sauces and fruit toppings, maple syrup, jams, jellies, biscuits, bread, pie and pizza dough, gelatine, coconut, vinegar, beef stew.

Sulphites are the most effective and cheapest antibrowning agents. Sulphites reduce quinones produced by catalysis of polyphenol oxidase to less reactive, and colourless compounds, thereby preventing pigmentation. The quantity of sulphites necessary to prevent enzymatic browning varies according to the concentration and the nature of the substrate, the level of activity exhibited by the polyphenoloxidase, the desired period of control and the presence of other inhibitors.

Besides anti-browning and anti-microbial functions, sulphites are used as dough conditioners in bakery industry, especially in frozen pizza dough and pie crusts. They are also used in biscuits, tortilla shells, cookies and crackers. The level of sulphites required for dough conditioners are relatively low. Sulphites act by breaking the cysteine disulphide bonds that are present in the gluten fraction of the dough. Cleavage of thiamine (vitamin B_1) in stored food, and destruction of cyanocobalamin (vitamin B_{12}) are undesired reactions of sulphites in foods.

Biological Data

Absorption, Distribution, Metabolism and Excretion

Sulphites are readily absorbed from the digestive tract. Normally, human serum has a sulphite concentration in the range 0–10 μM. Sulphite is oxidized in vivo to sulphate by sulphite oxidase, a molybdenum containing enzyme, located in the intra-membrane space of mitochondria. Sulphite oxidase is present in mammalian tissues and organs, mainly in liver, heart and kidney and very low amounts may be found in spleen, brain, skeletal muscle, and blood. In humans, sulphite oxidase activity varies from slightly to substantially less than in other mammals and some individuals present levels well below the normal range. During the oxidation of sulphite to sulphate in mitochondria, cytochrome c transfers the electrons produced to the electron transport chain, allowing generation of ATP in oxidative phosphorylation.

Metabolism of sulphites by sulphate oxidase has been shown to be rapid. Therefore, sulphite does not accumulate in the tissues on chronic administration. Sulphate, which represents the normal non-toxic sulphite metabolite, is excreted rapidly through the urine. In several species, less than 10% of the dose administered appeared in the urine as sulphite. A proportion of the sulphite absorbed is converted to thiosulfate. When elevated levels of thiosulfate in body fluids are observed it may be due to sulphite oxidase deficiency.

Adverse Reactions

Sulphites have been implicated as initiators of asthmatic reactions in a small subset of sensitive individuals of the asthmatic population. Adverse reactions, including anaphylactic reactions, dermatitis, urticaria, flushing, hypotension, abdominal pain and diarrhoea, have been reported in those individuals. The mechanism of action of sulphites in asthmatic reactions is not established, and various studies have proposed different possible mechanisms.

One study suggests that the reaction of nitric oxide and its carriers with sulphites may be a major source of sulphites' biological toxicity by affecting the pathway of the drugs used to treat asthma. Another study provided evidence that this response may be mediated through stimulation of an oro-bronchial reflex by sulphur dioxide vapour released from foods, rather than by an immunological mechanism. A third study says that partial sulphite oxidase deficiency is a possible mechanism involved in sulphite sensitivity due to alterations in drug metabolizing enzymes that can contribute to the varying susceptibility and response of sulphite-sensitive individuals to different drugs and/or therapeutics used for treatments.

Sensitivity to sulphites can develop at any time during a person's lifespan, with some initial reactions not showing up until a person has reached their forties or fifties. Dose-related respiratory hyper-reactivity has also been documented in a small number of individuals following consumption of potassium bisulphitetreated red wine and wine which contained smaller amounts of sulphites formed naturally as products of fermentation. The most sensitive patients have reacted to a dose of 5 mg sodium bisulphite.

On the other hand, gastric lesions had been induced in rats by sodium metabisulphite. Furthermore, gastro-intestinal reactions are also known to occur in man; with very high doses of sulphites, abdominal pain and vomiting have been observed in human volunteers.

Toxicological Data

Acute, Sub-chronic and Chronic Toxicity

The acute toxicity of free sulphites is low with a LD50 of 1000-2000 mg/kg bw. The acute toxicity of bound sulphites may even be lower with a LD50 that exceeds 5000 mg/kg bw researchers studied the toxicity of sulphite in newly weaned Wistar rats. 120 males and 120 females, were divided into six groups each of 20 males and 20 females. The groups were maintained on the stock diet containing 0.0 (control), 0.125, 0.25, 0.5, 1.0 or 2.0% sodium metabisulphite for periods up to 2 years and over three generations. A level of 2% sulphite caused slight growth retardation in two generation of rats, however it was no significant. Blood was present in the faeces in groups given 1% sulphite or more. Pathological examination revealed hyperplastic changes in both the fore- and glandular stomach with levels of 1 and 2% sulphite in each of the three generations. Also, it was perfomed a series of short-term studies in 10 male and 10 female rats on high sulphite levels (0–8%) for 10–56 days. Diets containing 6% sulphite caused considerable growth depression and reduced food intake and food efficiency. Anaemia occurred at 2% and above. Increase in weight of the spleen occurred with levels of 4% and above. The most sensitive criteria of sulphite damage in the present studies was the presence of blood in the faeces and changes in gastric morphology. The dietary level of sulphite that did not show any significant adverse effect was 0.25% in the diet, which is equivalent to 72 mg SO_2/kg of body weight after conversion and correction for sulphite losses.

he same research group examined sub-chronic and chronic oral toxicity of sodium metabisulphite. 20 male and 20 female pigs were fed with diets at levels of 0.0, 0.125, 0.25, 0.5, 1.0 or 2.0% of $Na_2S_2O_5$. After 15 weeks, 14 males and 14 females from each group were killed. The six males and six females remaining in each group were kept on the same diets for up to 48 weeks. The concentrations of sulphite still present in the various experimental diets when they were consumed by the pigs were calculated to be on average 0.06, 0.16, 0.35, 0.83 and 1.72% respectively. Organ-to-body weight ratios of the liver, kidneys, heart and spleen were significantly increased with dietary levels of 1.72%, after both 15 and 48 weeks; and with dietary levels of 0.83% only after 48 weeks. The no-effect level established in this study was 0.35% $Na_2S_2O_5$ in the diet of pigs for 48 weeks.

More recently, researchers determined the sodium metabisulphite sub-chronic toxicity on 24 female Wistar rats divided in four groups treated with 0.0, 0.25, 1 and 4% of sodium metabisulphite in their drinking water for 90 days. Administration of $Na_2S_2O_5$ at 1% and 4% had significant effects on body weight, food and water consumed. There was an increase in biochemical parameters (calcium, urea, creatinine, uric acid, transaminases) and decrease immunoglobulin levels. The hematology revealed a decrease in red blood cells and hemoglobin, as well as leucocytosis. They concluded that sub-chronic intake of $Na_2S_2O_5$ 1% and 4% seems to alter immune function, biochemical, hematological and physiological parameters in Wistar rats.

Genotoxicity and Carcinogenicity

Sulphites are capable of inducing mutations in yeast, E. coli and Vicia faba, but only at high concentrations (0.2 – 3 M) and pH values in the range of 5 – 5.6. Hayatsu and Miura described the mutagenic activity of sodium bisulphite on E. coli K12 after a treatment with a solution of 3 M $NaHSO_3$ and pH of 5.6 during 1.5 hours. Sodium bisulphite caused mutations by modification of cytosine. Similarly, researchers showed that a concentration of 1 M sodium bisulphite solution at pH 5.2 in E. coliK12is specific for cytosine-guanine mutants.

In a study in vitro, sodium bisulphite at low concentrations (0.05, 0.1, 0.5, 1.0, and 2.0x10^{-3} M) induced chromatid-type aberrations, but not chromosome-type aberrations in human blood lymphocytes; high concentrations induced both chromatid-and chromosome-type aberrations. The authors concluded that sulphur dioxide is a clastogenic and genotoxic agent.

Researchers however, did not find any carcinogenic effect in Wistar rats on the stock diet containing 0.0 (control), 0.125, 0.25, 0.5, 1.0 or 2.0% sodium metabisulphite for periods up to 2 years and over three generations.

Researchers examined the carcinogenicity of potassium metabisulphite in ICR/JCL mice. Three groups of 50 male and 50 female mice each received 0, 1, and 2% potassium metabisulphite solution in distilled water for 24 months. There was no significant difference in tumor incidence between the two experimental groups and the control group, suggesting that potassium metabisulphite is not carcinogenic in mice.

Sulphite has been shown genotoxicity in vitro but not in vivo. Furthermore, sulphites has a number of reactions with nucleic acids, and cellular molecules, but no harm has been yet been demonstrated in laboratory animals, generally rats or mice, to sulphite levels of 50 or 100 times in excess of the usual maximum human environmental exposure.

Acceptable Daily Intakes (ADI's)

The Joint FAO/WHO Expert Committee on Food Additives (JECFA) established an ADI of 0 –0.7 mg/kg-bw in 1974, which was maintained in its latest evaluation in the 51st report about evaluation of certain food additives. This ADI is for use with all the sulphiting agents presented in Table. The ADI was based on long-term studies in rats with a NOEL of 2.5 g/kg (0.25%) sodium metabisulphite in the diet, equivalent to 70 mg/kg of body weight per day of sulphur dioxide equivalents.

In the EU, the Scientific Committee for Food (SCF) in its 35th report established an ADI of 0 – 0.7 mg/kg of body weight in accordance to the established by the JECFA. Surveys on sulphite content in food and intake, carried out in different countries, showed that the acceptable daily intake (ADI) for this additive is likely to be exceeded. Nonetheless, a study performed in Belgium found that the intake of sulphites in the Belgian adult population is likely to be below the acceptable daily intake (ADI) with the possible exception for some high consumers of wines with high sulphite levels. Further, it was shown that for several food groups, actual use levels of sulphites are much lower than the maximum levels set in legislation.

Potassium Sorbate

Potassium sorbate (K-sorbate) is a food preservative commonly used in the baking industry to prevent mold, yeast, and microbes. It is often used in cakes and icings, beverage syrups, cheese, dried fruits, margarine, pie fillings, wine, etc. at concentrations dependent on the specific application.

It is a water soluble ingredient with molecular formula, $C_6H_7KO_2$. K-sorbate is commercially available in the form of powder or pellets. It is effective at pH up to 6 but drops rapidly at higher levels.

Potassium sorbate is produced by combining potassium hydroxide and sorbic acid to create a potassium salt. Sorbic acid is naturally present in the lactone form in berries such as rowan berries, Sorbus aucuparia L, which it was first isolated from. Some fruits such as cranberries, currants, strawberries naturally contain sorbic acid.

Commercial Production

Sorbic acid is commercially produced using the ketene–crotonaldehyde condensation method. It is purified by treating sorbic acid with sodium hydroxide, hydrochloric acid and activated carbon. The potassium salt can be produced from batch or sorbic acid production streams prior to drying. It is further granulated by extrusion and palletization.

Function

Similar to other sorbates, K-sorbate can:

- Inhibit microbial growth by changing the cell membrane morphology and integrity.
- Disrupt the transport functions and metabolic activity.
- Be more effective than other preservatives, such as calcium propionate and sodium benzoate in inhibiting mold growth in bakery products.
- Increase the product shelf life with limited impact on food quality. If used at very high concentration, it can have an undesirable effect on taste and flavor.

Application

K-sorbate is typically used in chemically-leavened products (dry blended with the flour) at a level of 0.03% to 0.4% of the batter weight. Because of its deteriorative effect on yeast cells, K-sorbate can reduce loaf volume and generate sticky dough that is difficult to process, therefore, it is not suitable for bread baking. K-sorbate can also be sprayed onto product surfaces after baking such as the case with tortillas.

References

- How-is-sodium-benzoate-used-in-food-524480: livestrong.com, Retrieved 10, August 2020
- Nitrates-nitrites, corporate-publications-170614: efsa.europa.eu, Retrieved 12, May 2020
- Nitrates-and-Nitrites-in-meat-products-322879821: researchgate.net, Retrieved 08, July 2020
- Short-Review-of-Sulphites-as-Food-Additives-273518332: researchgate.net, Retrieved 22, March 2020
- Potassium-sorbate, ingredients: bakerpedia.com, Retrieved 07, January 2020

Chapter 5

Membrane Technology in Food Processing

The technique of concentrating and separating particles without using heat is known as membrane technology. It includes technologies such as reverse osmosis, ultrafiltration, microfiltration and nanofiltration. This chapter closely examines the key concepts of membrane technology to provide an extensive understanding of the subject.

Membrane processing is a technique that permits concentration and separation without the use of heat. Particles are separated on the basis of their molecular size and shape with the use of pressure and specially designed semi-permeable membranes. There are some fairly new developments in terms of commercial reality and is gaining readily in its applications.

- Proteins can be separated in whey for the production of whey protein concentrate (WPC).
- Milk can be concentrated prior to cheese making at the farm level.
- Apple juice and wine can be clarified.
- Waste treatment and product recovery is possible in edible oil, fat, potato, and fish processing.
- Fermentation broths can be clarified and separated.
- Whole egg and egg white ultrafiltration as a preconcentration prior to spray drying.

Principle of Operation

When a solution and water are separated by a semi-permeable membrane, the water will move into the solution to equilibrate the system. This is known as osmotic pressure. If a mechanical force is applied to exceed the osmotic pressure (up to 700 psi), the water is forced to move down the concentration gradient i.e. from low to high concentration. Permeate designates the liquid passing through the membrane, and retentate (concentrate) designates the fraction not passing through the membrane.

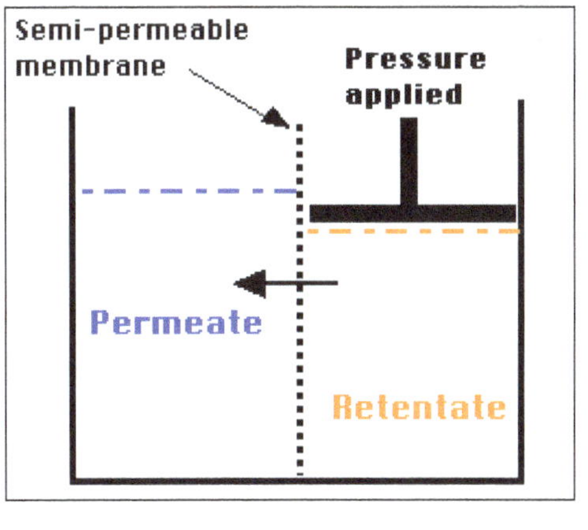

Membrane Separation Classification

Membrane separation processes are classified in to two: physical processes and chemical processes. Chemical processes are: hemodialysis, liquid membranes, active transport, supported membranes and facilitated transport. The physical membrane separation processes are: pressure driven process (microfiltration MF, ultra filtration UF, Nano filtration NF, reverse osmosis RO, gas separation), diffusional process (evaporation, perstraction, dialysis, membrane extraction, membrane absorption), heat process(membrane distillation, vacuum membrane distillation) and electric processes (electro dialysis, electrostatic pseudo liquid membrane).The key membrane technologies in the food industry are the pressure-driven membrane processes such as microfiltration (MF), ultra filtration (UF), Nano filtration (NF), and reverse osmosis (RO).Four membrane processes in their ascending pore diameter are Reverse osmosis (RO), Nano filtration (NF), Ultra filtration (UF) and Microfiltration (MF) is mainly used in food processing. UF systems as well as UF membranes have the maximum share (35%) in global membranes market followed by the share of MF systems and membranes (33%) and NF/RO systems and membranes (30%). Membrane contactors (MC), Electro dialysis (ED) and Pervaporation (PV) are the other membrane processes but contribute only for 2% share in total membranes market.

Membrane Generations

Membrane materials can be classified in four generations. First generation: Cellulose acetate derived membranes, which were developed to desalinate seawater. They were susceptible to microorganisms and disinfectants as well as sensitive to pH (3-8) and (maximum 50°C) temperature. Second generation: Developed from synthetic polymers like Polysulfone or polyolefin derivatives), these are better than first generation as having better resistant towards hydrolysis, pH and temperatures, but have lower resistance to mechanical compacting. Third generation: Mineral membranes consisting ceramic material based on zirconium or alumina oxide are usually deposited on a graphite surface. These membranes have great mechanical strength, are chemically inert and can be operated on high pressures, wide pH range (0-14) and temperatures greater than 400°C. The main demerit of these membranes is their cost (expensive than other membrane materials), but the same is compensated by their long operational life. Fourth gener-ation: Hybrid process i.e. combination of conventional Electro dialysis and different pore sizes membranes.

Reverse Osmosis

Reverse osmosis is the process in which pressure is applied to overcome colligative property and osmotic pressure that is directed by a thermodynamic parameter and a chemical difference of a solvent.

This application is mainly applied in the production of potable water in water plants and in industries. The end result will be the solute. It happens when the pure solvent is allowed to follow to one end of the membrane thus allowing a solute to retain in a permissible side of a membrane. Reverse osmosis removes suspended and types of dissolved species from water including bacteria.

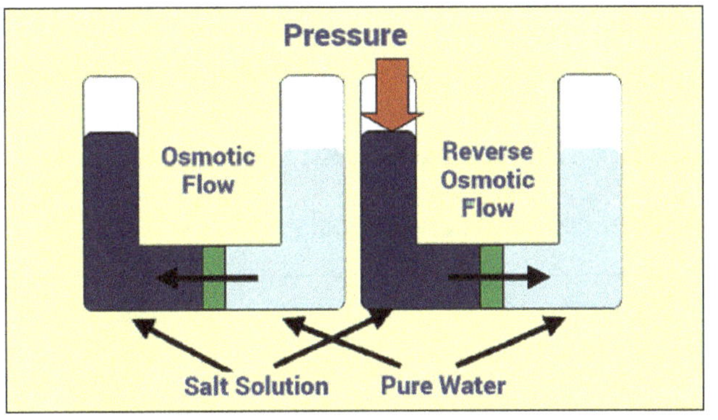

Reverse Osmosis Principle

Reverse osmosis works by reversing the principle of osmosis. The salt solution is subjected to pressure and pressed against the semi-permeable membrane. Here, the applied pressure is greater than the osmotic pressure. Thus, the molecules move from a highly concentrated solution to a less concentrated solution.

Working of Reverse Osmosis

- Diffusion is a process by which the molecules move from the region of higher concentration to lower concentration. There is a net movement meaning more molecules moving in one direction than in the opposite direction.
- In osmosis, the water molecules and the concentration gradient occurs over the semipermeable membrane which allows the entry of water and blocks the passage of ions and other larger molecules including sodium, chlorine, bacteria, glucose, etc.
- Reverse osmosis is the process or the technology which is used to remove ions, mineral chemicals, and other impurities from drinking water. In this process, greater pressure is applied, forcing the water to travel through the semipermeable membrane in opposite to natural osmosis.
- Reverse Osmosis works on the same principle as osmosis, but in the reverse direction. In this process direction of water flow is reversed by applying greater pressure.

For instance, consider a semipermeable membrane placed between the freshwater and concentrated aqueous solution. In natural osmosis, the freshwater will cross the semipermeable membrane and dilutes the concentrated solution. In reverse osmosis, the pressure is applied towards the concentrated aqueous solution and the water molecules are forced to cross the membrane towards the freshwater.

Contaminants Removed by Reverse Osmosis from Water

Reverse osmosis removes 99% of dissolved salts particles, colloids, bacteria, pyrogens from feed water. The contaminants are separated by the RO membrane on the basis of size and charge. The smaller the charge of the contaminant, the more are the chances for it to pass through the RO membrane. For eg., sodium and calcium are monovalent and divalent respectively. Due to their smaller charges, they can easily pass through the membrane. Similarly, RO cannot remove gases such as carbon dioxide from the water because they are not highly ionized.

Difference between Osmosis and Reverse Osmosis

Following are the major differences between osmosis and reverse osmosis:

Osmosis	Reverse Osmosis
This is the process by which the molecules of a solvent pass through the semi-permeable membrane from a region of lower concentration to a higher concentration.	This is the process by which the molecules of a solvent pass through the semi-permeable from a region of higher concentration to lower concentration when pressure greater than the osmotic pressure is applied.
It is a natural process.	It is an artificial process.
Occurs along the potential gradient.	Occurs against the potential gradient.
This is observed during the opening of stomata and absorption of water from the soil by the roots.	This is used in water purification systems.

Ultrafiltration

Ultrafiltration (UF) is basically a pressure-driven separation process, governed by a screening principle and dependent on particle size. Ultrafiltration membranes have a pore size between 1 nm and 100 nm, thus allowing retention of compounds with a molecular weight of 300 to 500 000 Dalton.

Typically, the process is suitable for retaining biomolecules, bacteria, viruses, polymers, colloidal particles and sugar molecules.

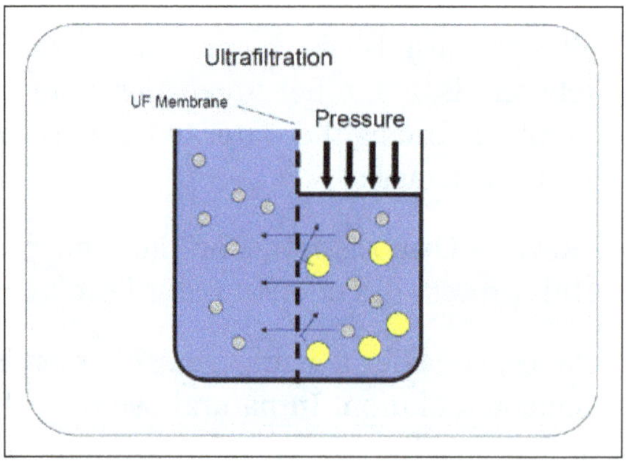

Ultrafiltration membranes are defined by their nominal molecular weight cut-off (MWCO). The MWCO generally represents the smallest molecular weight for which the membrane has a retention value of more than 90%. In many cases, however, separation efficiency is not only influenced by cut-off but also by interaction between the membrane and the raw solution. The operating pressure for ultrafiltration is usually between 0.1 and 1 MPa. The following operations are typical for the application of ultrafiltration:

- Filtration of raw solutions.

- Concentration of substances.
- Fractionating of substances.

In the fractionating of dissolved substances or the separation of dissolved substances and solvents, a sufficient degree of separation is reached when the size of the particles differs by the factor 10. Nowadays, different applications for ultrafiltration can be found in nearly all industrial sectors:

- Sterile filtration of drinking and beverage water.
- Treatment of surface water.
- Recovery of filter backflushing water.
- Separation of oil/water emulsions.
- Recovery of electrodeposition paint.
- Removal of metal hydroxides in wastewater treatment.
- Separation of biomass in biotechnology.
- Wastewater treatment and re-use.
- Membrane bioreactors.

Great significance is given to the use of ultrafiltration for sterile filtration because only a suitable ultrafiltration process can ensure a retention of >log 4 for bacteria, viruses, Legionella and possibly even endotoxins. Ultrafiltration is, therefore, the choice process for point-of-use filtration for the safe production of drinking water.

One of the most sophisticated uses of ultrafiltration lies in the application of membrane bioreactors (MBR) for wastewater. The ultrafiltration can operate in the normal way on a cross-flow bypass system, or submerged in the bioreactor vessel by means of vacuum suction at low trans-membrane pressures.

The combination of activated sludge with membrane separation in the MBR results in efficiencies of footprint, effluent quality and residuals production that cannot be attained when these same processes are operated in sequence.

Microfiltration

Micro-filtration (or MF for short) is one of the pressure-driven membrane processes in the series micro-filtration, ultra-filtration (UF), nano-filtration (NF) and reverse osmosis (RO). The micro-filtration process uses a membrane – a simple permeable material – which, in the case of micro-filtration, only allows particles smaller than 0.1 microns to pass through it. The micro-filtration membrane can consist of various materials like, for example, polysulfone, polyvinyldifluoride (PVDF), polyethersulfone (PES), ZrO_2 and carbon. The pore size varies between 0.1 and 5 microns. Because the pores are large compared to other mentioned filtration techniques, pressure – needed to send the liquid through a micro-filter membrane – is limited to 0.1 to 3 bar.

Micro-filter membranes are offered in various configurations by suppliers, with each configuration having a specific use and accompanying advantages and disadvantages. Possible membrane configurations include:

- Pipe-shaped membranes: capillary, hollow fibre or tubular.
- Plate-shaped membranes: flat plate or spiral.

In addition to the specific membrane configurations, one can also identify a few set-ups. The 2 most commonly used methods are dead-end and cross-flow set-ups. The names refer to the way in which the supply is sent to the membrane. In dead-end MF, the supply is sent directly to the membrane. The pollution layer will thus form on the supply side of the membrane surface. This layer contains all particles that have been separated on the basis of their size (sieve effect). This layer is periodically rinsed away by briefly re-sending the produced liquid (permeate) through the membrane in an opposite direction to the production flow. This helps to loosen the hardened layer, and makes it ready for disposal. This is referred to as a semi dead-end set-up.

This re-rinsing may not be enough to remove the layer from the surface if the hardened layer is too strongly compressed or if the bond with the membrane is very strong. In this case, chemical cleaning must be implemented with, for example, bleach, peroxide, acid and alkali or detergent. The figure below shows a dead-end set-up, where M = the membrane.

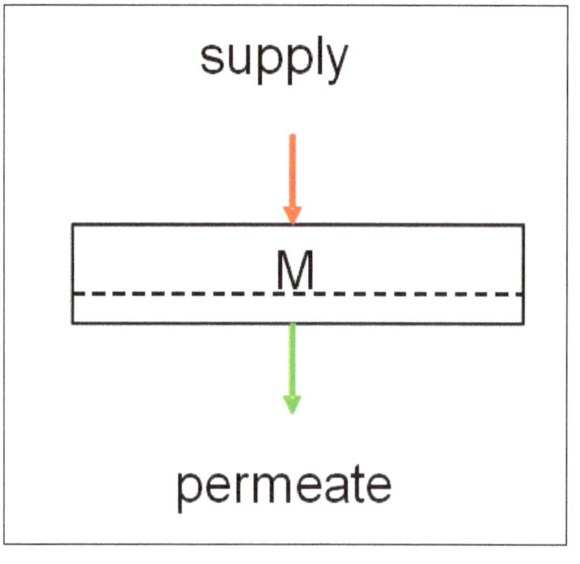

In a cross-flow set-up, the liquid is passed along the membrane surface at a particular speed. Them permeate is able to pass through the membrane and the larger particles are left behind in a concentrated flow (the retentate). The hardened layer in this set-up is continuously removed by the cross-flowing supply flow.

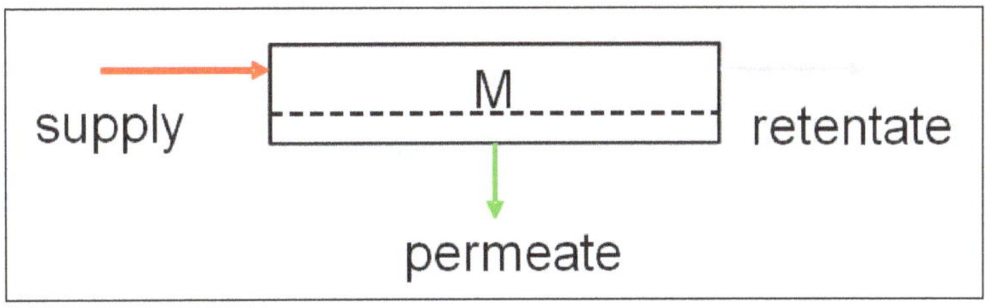

Specific Advantages and Disadvantages

The advantages of MF are:

- Low operating pressure required.
- Low energy consumption for semi dead-end set-up, compared to nano-filtration or reverse osmosis.
- Few manual actions required.
- Relatively cheap.
- No energy-consuming phase transfer needed, such as e.g. evaporation techniques.
- Quality of the produced permeate is not determined by the management.

Possible disadvantages of MF are:

- Only suspended matter and bacteria removed ((~log 5 removal).
- Sensitive to oxidative chemicals (e.g. nitric acid, sulphuric acid, peroxide and persulphate in high concentrations).
- Damage can be caused by hard and sharp particles > 0.1 mm, whereby pre-filtration is necessary.
- Membrane damage if re-rinsed at pressure in excess of 1 bar.

Application

Micro-filtration is primarily used as a pre-treatment step in the production of drinking and process water. It has excellent properties for removing suspended matter, bacteria and cysts. It is an alternative to classic sand filtration. Further, cross-flow MF is used in the:

- Dairy industry (cheese, milk,).
- Food industry (clarification of fruit juice, wine, beer, etc).
- Metal industry (oil/water emulsion separation).
- Textile industry (effluent treatment).
- Pharmaceutical industry (cold sterilisation).

MF is also implemented for the re-use of wastewater. In this case, it can be combined with a biological step in a membrane bioreactor.

Boundary Conditions

Membranes must be protected against hard particles larger than 0.1 mm. They can be removed by regular pre-filtration. Further, supply flows and pH conditions must be compatible with the membrane material.

Effectiveness

MF can be implemented for removing the following parameters:

- Suspended matter (>99%).
- Harmful micro-organisms (e.g. bacteria, protozoa, algae, fungi) (>99%).
- PAHs.

Further, MF can also be implemented to break down emulsions.

Nanofiltration

Nanofiltration is sometimes used to recycle wastewater, as it offers higher flux rates and uses less energy than a reverse osmosis system. The design and operation of nanofiltration is very similar to that of nanofiltration is very similar to that of reverse osmosis, with some differences. The major difference is that the nano membrane is not as "tight" as the reverse osmosis membrane. It operates at a lower feedwater pressure and it does not remove monovalent (i.e., those with a single charge or valence of one) ions from the water as effectively as the RO membrane.

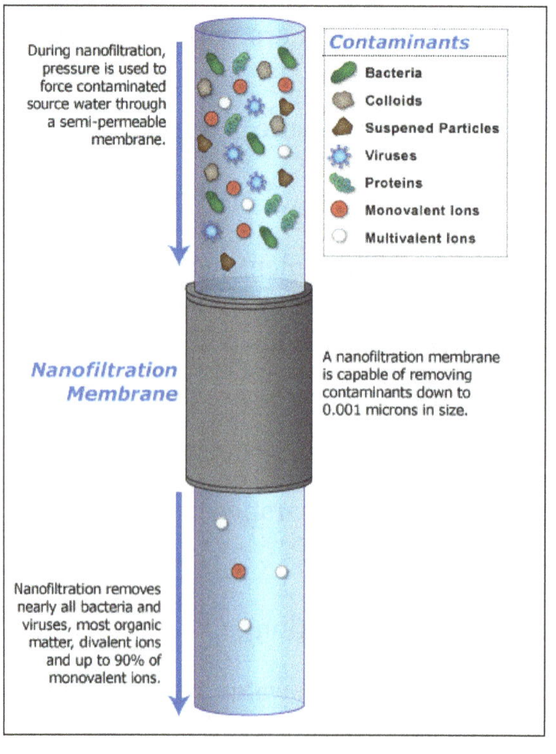

While a RO membrane will typically remove 98-99% of monovalent ions, such as chlorides or sodium, a nanofiltration membrane typically removes 50% to 90%, depending on the material and manufacture of the membrane. Because of its ability to effectively remove di and trivalent ions, nanofiltration is frequently used to remove hardness from water while leaving the total dissolved solids content much less affected than would RO. For this reason, it has been called the "softening membrane". Nanofiltration is often used to filter water with low amounts of total dissolved solids, to remove organic matter and soften water.

Because it is a "looser membrane", nanofiltration membranes are less likely to foul or scale and require less pretreatment than reverse osmosis systems. Sometimes it is even used as pre-treatment for reverse osmosis. Nanofiltration can be employed in a variety of water and wastewater treatment industries for the cost-effective removal of ions and organic substances.

Besides water treatment, nanofiltration is often employed in the manufacturing process for pharmaceuticals, dairy products, textiles and bakeries.

References

- Reverse-osmosis, biology: byjus.com, Retrieved 18, February 2020
- Membrane-processes: fumatech.com, Retrieved 06, July 2020
- Microfiltration: vito.be, Retrieved 30, January 2020
- Nanofiltration: waterprofessionals.com, Retrieved 14, March 2020

Permissions

All chapters in this book are published with permission under the Creative Commons Attribution Share Alike License or equivalent. Every chapter published in this book has been scrutinized by our experts. Their significance has been extensively debated. The topics covered herein carry significant information for a comprehensive understanding. They may even be implemented as practical applications or may be referred to as a beginning point for further studies.

We would like to thank the editorial team for lending their expertise to make the book truly unique. They have played a crucial role in the development of this book. Without their invaluable contributions this book wouldn't have been possible. They have made vital efforts to compile up to date information on the varied aspects of this subject to make this book a valuable addition to the collection of many professionals and students.

This book was conceptualized with the vision of imparting up-to-date and integrated information in this field. To ensure the same, a matchless editorial board was set up. Every individual on the board went through rigorous rounds of assessment to prove their worth. After which they invested a large part of their time researching and compiling the most relevant data for our readers.

The editorial board has been involved in producing this book since its inception. They have spent rigorous hours researching and exploring the diverse topics which have resulted in the successful publishing of this book. They have passed on their knowledge of decades through this book. To expedite this challenging task, the publisher supported the team at every step. A small team of assistant editors was also appointed to further simplify the editing procedure and attain best results for the readers.

Apart from the editorial board, the designing team has also invested a significant amount of their time in understanding the subject and creating the most relevant covers. They scrutinized every image to scout for the most suitable representation of the subject and create an appropriate cover for the book.

The publishing team has been an ardent support to the editorial, designing and production team. Their endless efforts to recruit the best for this project, has resulted in the accomplishment of this book. They are a veteran in the field of academics and their pool of knowledge is as vast as their experience in printing. Their expertise and guidance has proved useful at every step. Their uncompromising quality standards have made this book an exceptional effort. Their encouragement from time to time has been an inspiration for everyone.

The publisher and the editorial board hope that this book will prove to be a valuable piece of knowledge for students, practitioners and scholars across the globe.

Index

www.ingramcontent.com/pod-product-compliance
Lightning Source LLC
LaVergne TN
LVHW080913210926
848485LV00004B/647